Are We Alone?

B.J. Mochejska (CfA), J. Kaluzny (CAMK), 1m Swope Telescope

Are We Alone?

HUMANKIND'S SEARCH FOR EXTRATERRESTRIAL CIVILIZATIONS

• • •

Kaushik Sengupta

© 2017 Kaushik Sengupta
All rights reserved.

ISBN: 197654999X
ISBN-13: 9781976549991
Library of Congress Control Number: 2017914664
CreateSpace Independent Publishing Platform
North Charleston, South Carolina

KAUSHIK SENGUPTA

Are We Alone?

• • •

KAUSHIK SENGUPTA IS AN EXPERIMENTAL particle physicist by training and a software engineer by profession. While he is passionate about both, Kaushik particularly enjoys bringing academic science to popular audiences. He began his career as a research physicist at State University of New York (SUNY) Buffalo and Louisiana State University (LSU), collaborating with various international and NASA scientists on high-energy particle physics experiments, publishing his results extensively. Early in this professional life, he became interested in mathematical computing and software design. He teamed up with AT&T engineers to design software packages on data modeling and statistical forecasting. He also taught physics at the university level in India. Kaushik lives in Kendall Park, New Jersey, with his wife and daughter. This is his first book. You can follow his thoughts on cosmicglimpses.blog.

Contents

Preface ix
Introduction xiii
    The History of the Search for Extraterrestrial Life ... xiv

Chapter 1 The Search 1
    The Wow! Signal 3
    A False Signal 6
Chapter 2 The Dyson Sphere 9
Chapter 3 The Ranking of Civilizations 13
    The Drake Equation 13
    The Drake Equation Revisited 16
    The Fermi Paradox 17
    Kardashev Civilizations 20
Chapter 4 The Great SETI Debate 23
Chapter 5 The Search for Alien Life in Our Solar System 29
    Enigmatic Europa 30
    The Ocean World of Enceladus 33
    Organic Material on Ceres 35
    What Does All This Mean? 37
Chapter 6 Exoplanets 38
    Seager's Equation 42
    The Seven Worlds of TRAPPIST-1 44
    Interstellar Travel 50
    Exoplanets and SETI 52

Chapter 7	The Current State of SETI	54
	Project SERENDIP	55
	SETI@home	57
	A Message from Earth	59
Chapter 8	Hints?	63
	Boyajian's Star	64
	Fast Radio Bursts (FRBs)	66
Chapter 9	What Next?	70
	Further Reading	73
	Appendix A	75
	How Exoplanets Are Detected	75
	Appendix B	79
	New Telescopes	79
	Transiting Exoplanet Survey Satellite	79
	James Webb Space Telescope	80
	Appendix C	83
	The Twenty-One-Centimeter Hydrogen Line	83
	References	85
	Index	91

Preface

• • •

Haven't we all looked at the sky full of stars on a clear, cloudless night and wondered with awe and inspiration if there were other living beings just like us in a distant galaxy far, far away?

My inspiration came one night in 1984 during a trip to the Sundarbans on the Ganges Delta, when I looked up at the night sky from the roof of my boat. It was pitch black everywhere except the sky, which was laced with millions of stars. For the first time in my life, I could actually see the Milky Way—the dim glowing band of light arching across the night sky. I was utterly unfamiliar with the spectacle, as I was raised in the sprawling suburbs of a big city where the light pollution caused all but the brightest of stars and planets to be masked, even on a clear night. But in this remote region of the Sundarbans, far from civilization, the vastness of the universe grew on me. I was dumbfounded with the beauty of the star-studded sky, and I wanted to know everything about stars—how they were formed, the galaxies in which they reside, and the nature of space, time, and gravity. Although I was pursuing a research career in experimental particle physics at the time, I became fascinated with astrophysics because it promises to open a window through which one can see farther into the universe.

Astrophysics became my hobby. I started following the latest research, eagerly reading the popular books written by experts in the field. It was not so much the physics that influenced me; I became enamored of the realization that as a species, we are obsessed with ourselves. We are engrossed in our own politics, our history, our philosophy, and our gods—as if we are the center of the universe. In fact, we are inhabitants of a fragile planet surrounded by a wafer-thin strip of atmosphere (relatively speaking); three-quarters of its surface is filled with water, the main ingredient that makes life possible. The planet that we call home revolves around an average star, the Sun, which is located in the inner rim of a minor spiral arm of the Milky Way galaxy called the Orion Arm. There are hundreds of billions of stars in the Milky Way, and a hundred billion galaxies in the observable universe; the majority of them, if not all, are thought to be orbited by planets. The question that naturally comes to mind is, "Are we alone in this universe?"

A commonly used definition of life is "a self-sustaining system capable of Darwinian evolution." Some scientists believe that we will never fully understand what life on Earth actually is until an alternative is found elsewhere, based on the "same DNA, metabolism, and carbon-based blueprint shared by all known life on Earth" (Kaufman n.d.). Life could be as simple as a single-celled organism, or it could even be the hypothetical beings of science-fiction literature that descend from civilizations far more advanced than our own. Finding irrefutable evidence of alien life, no matter how elementary in form, would undeniably represent a fundamental revolution in science; we would enter a brand-new phase of human civilization that would force us to redefine our place in the universe. While humankind may always be interested in its own laws and societies, alien life could force us to reckon with our own insignificance and ultimately help us know ourselves better. This question—whether we are the only intelligent species in our galaxy, or in the entire universe, for that matter—is one of the most important philosophical questions that science must answer. My goal in this book is to present some of the recent scientific advances made in an effort to answer this question.

Finding evidence of a cosmic civilization is challenging, because scientists do not know what exactly its signature would be. But ingenious and educated guesses have been made by scientists over the years—you will find all those stories in this book. What features of the universe are essential for the emergence of intelligent beings such as ourselves? Is it merely coincidental, or is there a deeper scientific reason? If intelligent life does exist beyond Earth after all, is there a way of communicating with them? Is interstellar travel feasible? Finally, what is the future of life on Earth and beyond? These are some of the most fascinating questions that face humanity and the subject of this book.

This book is for somebody with a curious bent of mind, not necessarily with a background in physics, although some exposure to undergraduate physics will definitely be beneficial—somebody looking to be told an engaging and intellectually stimulating story who will not feel patronized if a few difficult ideas are explained to make a point. This is the story of an intellectual adventure that is still unfurling, the quest for the knowledge that has eluded us so far, but with the hope of finding an answer perhaps within our lifetimes.

Kaushik Sengupta
Kendall Park, New Jersey

Introduction

• • •

*Sometimes I think we're alone in the universe, and sometimes
I think we're not. In either case the idea is quite staggering.*

—Attributed to Arthur C. Clarke

On December 16, 2014, at the fall meeting of the American Geophysical Union in San Francisco, scientists announced that NASA's *Curiosity* rover had detected sudden spikes of methane in the Martian atmosphere. The scientists also confirmed for the first time the presence of carbon-based organic molecules in a Martian rock sample.

Why is this newsworthy?

Because this could indicate that a carbon-based life-form could have existed on the Red Planet at some point during its cosmic evolution. This is also a clue that Mars may currently harbor life, possibly in a microbial form. Life as we know it on Earth produces significant amounts of methane, thus the spikes in methane in the Martian atmosphere may signal a similar form of life on the planet. However, since methane can also be produced by geological means, the detection of methane is not directly indicative of signs of past or present life-forms. Rather, it suggests the possibility that Mars once had the ingredients required for life and might still harbor them—one only needs to know where to look. The origin of Mars's methane has become an active area of research, with missions such as *Curiosity* and India's *Mars Orbiter* keenly measuring the changes in its abundance.

On the other hand, the answer to the question of whether intelligent life exists somewhere in the universe is still unknown. Humankind has long been intrigued by the idea of whether life on Earth is unique in the vastness of space, or whether our galaxy is swarming with extraterrestrial civilizations.

Today, most scientists believe that advanced life-forms exist somewhere else in the universe. In a documentary series, the famous British astrophysicist Stephen Hawking argued that it is "perfectly rational" to assume intelligent life exists elsewhere in the universe. Hawking believes that primitive life is very common, but intelligent life is probably fairly rare.

The History of the Search for Extraterrestrial Life

> Is ET out there? Well, I work at the SETI Institute. That's almost my name. SETI: Search for Extraterrestrial Intelligence. In other words, I look for aliens, and when I tell people that at a cocktail party, they usually look at me with a mildly incredulous look on their face. I try to keep my own face somewhat dispassionate.
>
> —Seth Shostak

Throughout history, humans have gazed up at the sky in awe and wonder. Our ancestors recorded the complex movements of stars, identified constellations, and marked time by creating celestial-based calendars. The stars also became the sources of countless myths and legends.

In the fifth century BC, the ancient Greek philosopher Democritus imagined "innumerable worlds of different sizes," some brimming with life. The Roman poet Titus Lucretius Carus wrote about "other worlds" with "different tribes of men, kinds of wild beasts."

The science-fiction genre began in the seventeenth century with the great German astronomer Johannes Kepler writing about a voyage to the Moon during which travelers encountered reptilelike creatures. Later in that century, the Dutch mathematician Christiaan Huygens concluded that some of the other planets must harbor life and speculated on the conditions there.

In 1894, while viewing the Red Planet with his telescope, the American astronomer Percival Lowell spotted a web of "canals" on the surface of Mars. The structures appeared so elaborate to him, they could have been built only by intelligent entities.

During the last century, NASA and other space agencies around the world started sending satellites into space to explore our solar system and to take photographs of the planets and their moons. Robotic devices were sent in unmanned spacecraft to explore their surfaces.

Figure 0-1: Picture of an anonymous wood engraving called the Flammarion woodcut. It first appeared in 1888 in L'atmosphere: meteorolgie populaire (The Atmosphere: Popular Meteorology) by Camille Flammarion (1842–1925), a French astronomer. The engraving appears to depict a pilgrim looking past the visible sky at the hidden structure of the universe.

In 1969, the first astronauts walked on the Moon and brought back lunar rocks and dust. Evidence of water was found on the Moon and Mars, as well as on Jupiter's moon Europa. NASA's Hubble Space Telescope recently detected faint signatures of water in the atmospheres of five planets beyond our solar system. Amino acids, the building blocks of life, were discovered in meteorites that have fallen to Earth. In November 2014, the European Space Agency's Rosetta mission successfully placed a small spacecraft on the surface of a speeding comet 67P/Churyumov-Gerasimenko. And recently, NASA discovered significant spikes in methane levels on our neighboring Mars. Moreover, new and more powerful telescopes and better analytical tools have led to the discovery of hundreds of new planets orbiting other stars.

The search for extraterrestrial intelligence is still in its infancy, and no one has found life anywhere other than on Earth. However, the sheer immensity of our own galaxy and the even greater immensity of our universe (estimates place the number of stars in the Milky Way at between two hundred and four hundred billion and the number of galaxies in the universe at one hundred billion) have led many to realize that Earthlike planets may be quite commonplace within our universe. Moreover, there may be a high chance of life on many of them—some of them may even harbor intelligent life.

Why should we care?

It can be argued that developing an understanding of our place in the universe is key to the survival of our species. That knowledge can tell us something about our future technological prospects and the existential risks confronting us and may even throw light on the nature of the evolution of the human species.

Unfortunately, the search for extraterrestrial intelligence (SETI) is not a government-funded effort. In fact, the prevailing belief in the US Congress is that SETI is a waste of taxpayer money since there are other, far-more pressing issues facing the nation today. Interestingly, SETI was once a project spearheaded by NASA, but the 103rd Congress eliminated the program thirty years ago. The same session saw the demise of the Superconducting Super Collider, and with that vanished America's hope of retaining her dominance in high-energy physics.

I will argue that SETI as a scientific discipline is no less important than other scientific areas that have no immediate payoffs or benefits. We have built powerful particle accelerators to understand the nature of matter; we have mapped life's genetic code with the long-term and hopeful notion of creating cutting-edge therapies to prolong and enrich lives. Yet none of these endeavors promise to produce immediate practical applications. SETI, after all, is a quest for knowledge, and the new tools developed in SETI research have the potential to advance our scientific know-how by stimulating the development of a host of technologies such as superior satellite technology, materials science applications, telescope technology, and so on. These advances are expected to invigorate public interest in scientific education, develop new skills, create new employment opportunities, and even stimulate the world economy. To quote Arthur C. Clarke, "SETI is probably the most important quest of our time, and it amazes me that governments and corporations are not supporting it sufficiently."

SETI research stems from the urge to explore the unknown and push the envelope of humankind's persistent quest for knowledge. Finding the answer to the age-old question of whether we are alone in our galaxy may well be within our technological reach, and some scientists believe it may even happen during our lifetime.

CHAPTER 1

The Search

• • •

The reader may seek to consign these speculations wholly to the domain of science-fiction. We submit rather, that the foregoing line of argument demonstrates that the presence of interstellar signals are entirely consistent with all we now know...We therefore feel that a discriminating search for signals deserves a considerable effort. The probability of success is difficult to estimate; but if we never search the chance of success is zero.

—Giuseppe Cocconi and Philip Morrison

In 1959, Giuseppe Cocconi and Philip Morrison of Cornell University published a landmark article in the journal *Nature* entitled "Searching for Interstellar Communications," which offered a realistic strategy for a search for extraterrestrial intelligence. The paper, now considered a classic, contained the following lines: "...near some star rather like the Sun there are civilizations with scientific interests and with technical possibilities much greater than those now available to us...We shall assume that long ago they established a channel of communication that would one day become known to us, and that they look forward patiently to the answering signals from the Sun, which would make known to them that a now society has entered the community of intelligence. What sort of a channel would it be?" (Cocconi and Morrison 1959, 844–46).

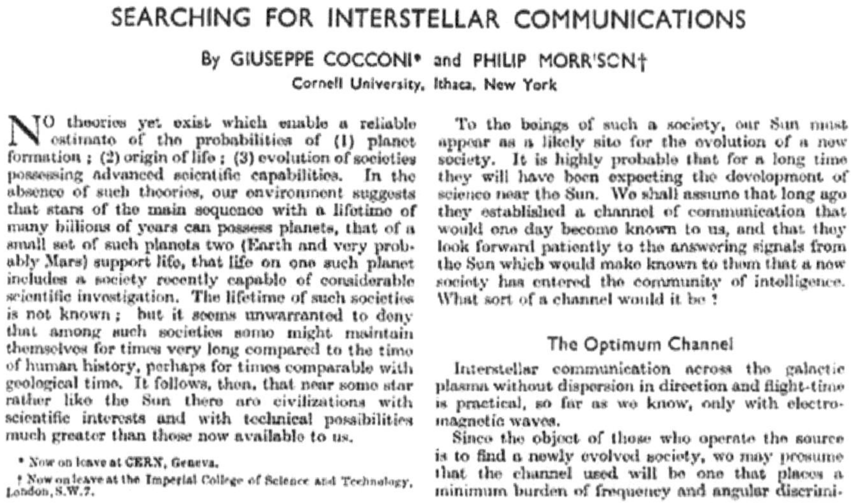

SEARCHING FOR INTERSTELLAR COMMUNICATIONS

By GIUSEPPE COCCONI* and PHILIP MORRISON†

Cornell University, Ithaca, New York

NO theories yet exist which enable a reliable estimate of the probabilities of (1) planet formation; (2) origin of life; (3) evolution of societies possessing advanced scientific capabilities. In the absence of such theories, our environment suggests that stars of the main sequence with a lifetime of many billions of years can possess planets, that of a small set of such planets two (Earth and very probably Mars) support life, that life on one such planet includes a society recently capable of considerable scientific investigation. The lifetime of such societies is not known; but it seems unwarranted to deny that among such societies some might maintain themselves for times very long compared to the time of human history, perhaps for times comparable with geological time. It follows, then, that near some star rather like the Sun there are civilizations with scientific interests and with technical possibilities much greater than those now available to us.

* Now on leave at CERN, Geneva.

† Now on leave at the Imperial College of Science and Technology, London, S.W.7.

To the beings of such a society, our Sun must appear as a likely site for the evolution of a new society. It is highly probable that for a long time they will have been expecting the development of science near the Sun. We shall assume that long ago they established a channel of communication that would one day become known to us, and that they look forward patiently to the answering signals from the Sun which would make known to them that a new society has entered the community of intelligence. What sort of a channel would it be?

The Optimum Channel

Interstellar communication across the galactic plasma without dispersion in direction and flight-time is practical, so far as we know, only with electromagnetic waves.

Since the object of those who operate the source is to find a newly evolved society, we may presume that the channel used will be one that places a minimum burden of frequency and angular discrimi-

Figure 1-1: Cocconi and Morrison's classic paper published in Nature, in which they were the first to formally argue a scientific rationale for SETI.

Cocconi and Morrison figured there would be one radio channel at a frequency of about 1,420 megahertz, the frequency of the hydrogen atom, the most abundant element in the universe. This universal frequency should be known to all civilizations capable of communication across space, so Cocconi and Morrison suggested that the frequency band close to 1,420 megahertz should be searched. (See appendix C.)

Cocconi's and Morrison's argument is still valid today. Additionally, they assumed the following:

- The majority of other life-forms would be based on the same chemistry as those on Earth, where life is based on the chemistry of carbon.
- Water is a necessity ingredient for life.

Cocconi's and Morrison's work transformed a speculative curiosity into a scientific discipline. Scientists could now conduct a search for an alien civilization by making use of the strategy they so brilliantly propounded.

And indeed, the first radio search was conducted one year later in 1960 by the radio astronomer Frank D. Drake of the National Radio Astronomy Observatory (NRAO) in Green Bank, West Virginia, using a single channel around that frequency.

Project Ozma,[1] as it was called, focused on two stars about the same age as our Sun and about eleven light-years (sixty-six trillion miles) away from us: Tau Ceti in the constellation Cetus (the whale) and Epsilon Eridani in the constellation Eridanus (the river).

Project Ozma's eighty-five-foot NRAO radio telescope was tuned to the twenty-one-centimeter emission (1,420 megahertz) coming from cold hydrogen gas in interstellar space. A single one-hundred-hertz channel receiver scanned four hundred kilohertz of bandwidth for six hours a day from April to July 1960. Drake's team of astronomers looked for a repetitive series of uniformly patterned pulses that would indicate a message sent by intelligent beings or a series of prime numbers such as one, two, three, five, or seven. But none was detected, with the exception of an early false alarm that the team was able to eliminate as being caused by a secret military experiment. The only sound that came from their equipment was random static.

Although Project Ozma did not find any alien civilizations, the pioneering steps it took more than fifty years ago enabled SETI to become a feasible scientific objective. For years various radio telescopes around the world were commissioned to survey the sky, searching for something that might arrive from an alien civilization. For years they heard nothing except white noise.

THE WOW! SIGNAL

Then one night in August 1977, a radio telescope in the United States picked up a remarkable signal from direction of the constellation Sagittarius.

---

[1] Drake was aware of the unprecedented and fantastical nature of his project. In his book *Is Anyone Out There?* (with Dava Sobel, 1992), he writes, "Even I, in my fever of enthusiasm, couldn't assume that we would really detect an intelligent signal." Drake called his radio search Project Ozma, naming it after the princess in a children's story entitled "Ozma of Oz."

A few days later, a volunteer researcher for Ohio State University's now-defunct Big Ear radio observatory named Jerry Ehman was flipping through the computer printouts generated by the telescope's scan of the skies on August 15. In those days, such in-

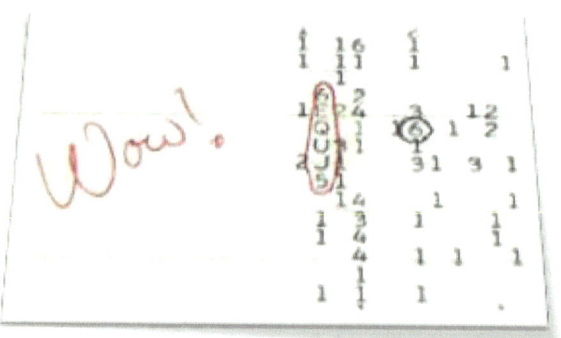

Figure 1-2: The Wow! Signal. Credit: Wikimedia Commons.

formation was run through an IBM 1130 mainframe computer, printed on perforated paper, and then painstakingly examined manually. Ehman spotted a vertical column with the alphanumerical sequence "6EQUJ5," which had occurred at 10:16 p.m. EST. Clearly astounded, he wrote *Wow!* on the original printout of the signal (figure 1-2).

The Wow! Signal, as it came to be known, was loud and clear and lasted for seventy-two seconds. It was defined by a string of letters—6EQUJ5—superimposed in a long sequence of low numbers—ones, twos, threes, and fours—which represented the background hum of an ordinary signal. Each digit on the printout represented the intensity of a radio signal from zero to thirty-five, with intensities above nine being represented by letters. Thus, *U* represented a signal that was thirty times higher than the background noise level. As the telescope swept across the sky, something extraordinary caused the signal to surge, and the computer started outputting letters instead of numbers. What perhaps is most interesting is that the frequency of the Wow! Signal resonated at 1,420 megahertz, right on the fabled hydrogen line, which precisely is the frequency band that Cocconi and Morrison thought could be used by an alien civilization for interstellar signaling.

It was possible that the Wow! Signal had terrestrial origin. But careful analysis by scientists eliminated that possibility.

The signal intensity had the expected shape of a typical narrowly focused bell curve—the intensity of the signal grew as Big Ear approached the signal and decreased as Big Ear moved away. (See figure 1-3.) This meant that the signal could have originated from a single point in space. Could the signal have been produced by a moving object, such as a satellite? That possibility was also eliminated for two reasons. First, only astronomical research is allowed to be conducted at the 1,420 megahertz frequency (the protected spectrum), so it is extremely unlikely that the Wow! Signal was man made. Second, to duplicate the signal, the satellite would have to have been located at just the right place and moving precisely in line with Big Ear, which is highly unlikely as well.

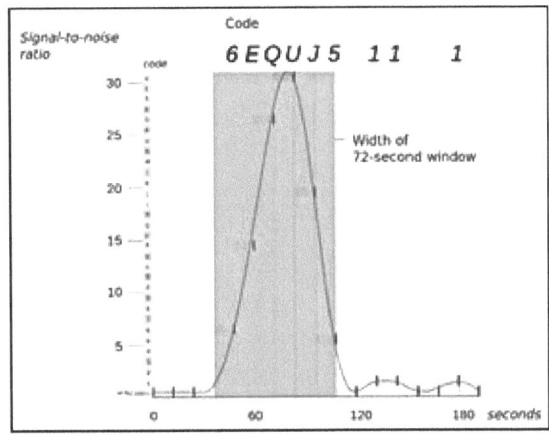

Figure 1-3: The Wow! Signal had the bell curve of intensity expected of a true space signal recorded by Big Ear.

The most significant fact was the signal's frequency being so close to the hydrogen line. This, combined with the signal's intensity distribution over the seventy-two seconds it was recorded, supported the conclusion that the signal originated somewhere near the Chi Sagittarii star group. Interestingly, if the signal indeed came from an alien civilization, it would have required a 2.2-gigawatt transmitter, vastly more powerful than any existing terrestrial radio station.

A search was immediately conducted for a repeat of the Wow! Signal. Scientists scanned the sky in the direction of the constellation Sagittarius, where the signal had originated. More and more sensitive telescopes were used, along with improved software that was designed to find signals among the background noise.

Despite several decades of searching, the signal has never been found again. The Wow! Signal remains the only reliably recorded signal apparently received from deep space that has the quality of an intentional signal. Where the signal came from is still a mystery.

We may never know if the Wow! Signal was a message from an alien species or a radio burst from a natural source that coincidentally hit the hydrogen line.

A False Signal

The history of SETI research is full of false positives.

This section tells you the story of a major discovery in the field of astronomy when interestingly, the researchers initially could not rule out the possibility that their signal could have been sent by intelligent beings from another world. Eventually, it proved to be a false SETI signal. But the research led to a PhD and a Nobel Prize in physics.

In 1967, Antony Hewish, a radio astronomer at Cambridge University, was designing a radio telescope to observe the scintillation (twinkling) of stars, particularly quasars.[2] Stars appear to scintillate or twinkle because the light has to pass through Earth's atmosphere to reach an observer on Earth. Refraction of starlight through the various layers of the atmosphere, which are at different temperatures and densities, cause them to twinkle or scintillate. Luckily, the atmosphere is transparent to radio waves, so Hewish realized radio telescopes could be built around Cambridge to study his quasars!

Hewish's graduate student Jocelyn Bell was responsible for operating the radio telescope and analyzing the data. In the summer of 1967, Bell

---

[2] Quasars or quasi-stellar radio sources, first detected in the 1960s, are a class of celestial objects called active galactic nuclei (AGN). They are the most distant and most luminous objects in the universe, and they are powered by a supermassive black hole located in the center of a galaxy. While the nature of these objects was controversial until the early 1980s, there is now a scientific consensus that a quasar is a compact region in the center of a massive galaxy surrounding a central supermassive black hole. Today, we know most quasars are faint radio emitters. In addition to radio waves and visible light, quasars also emit ultraviolet rays, infrared waves, x-rays, and gamma rays. Most quasars are larger than our solar system.

observed an unusual signal at a wavelength of 3.7 meters, "scruff" as she called it, which manifested as sharp bursts of radio energy at a regular interval of 1.3 seconds, much faster than any known stellar rotation rates.

After extensive analysis, Hewish and Bell realized the pulses were arriving from outside our solar system. The duration of a pulse was only sixteen milliseconds. This short duration suggested that the source could be no larger than a small planet.

But if another planet was the source of these radio waves, as that planet orbited its alien sun, the frequency of the radio waves would be higher when the planet moved toward us and lower when it moved away. This Doppler shift in frequency is the reason for the apparent change in pitch of a car or train as it races past an observer. But Hewish's subsequent careful measurements indicated the change in the frequency could be explained as a result of the motion of Earth (the observer) around our Sun. Thus, the source of the radio waves could not have been a planet.

Figure 1-4: Jocelyn Bell photographed in 1968 outside the Mullard Radio Astronomy Observatory at the University of Cambridge. Image credit: National Media Museum/ Science & Society Picture Library.

Interestingly, Bell and Hewish first thought the signal could possibly be a beacon from an alien source and labeled it LGM for "little green men." This may sound rather amusing today, but imagine yourself in the shoes of Jocelyn Bell and her adviser Antony Hewish! In Bell's own words, "Here was I trying to get a PhD out of a new technique, and some silly lot of little green men had to choose my aerial and my frequency to communicate with us" (Bell 1977, 685–89). Soon Bell found another pulsing signal in a completely different part of the sky, slightly faster, with a pulse rate 1.2 per second. This meant the pulses must have a less exotic explanation.

"It was very unlikely that two lots of little green men would both choose the same, improbable frequency, and the same time, to try signaling to the same planet Earth" (Bell 1977, 685–89).

With insights from theoretical astrophysicists, Hewish determined that the regularly patterned radio signals, or pulses, that Bell had detected were not caused by Earthly interference or by any intelligent life-forms trying to communicate with distant planets, but rather they were energy emissions from certain stars called pulsars. Fred Hoyle suggested that the pulsar could be the remnants of a supernova explosion. Thomas Gold of Cornell University in Ithaca, New York, suggested that the signals observed by Jocelyn Bell were coming from neutron stars[3], but the neutron stars were spinning around an axis. A neutron star wouldn't need to be emitting pulses of radiation but could emit a steady radio signal if it swept around in circles like light from a lighthouse. When the pulsar "lighthouse" was pointing toward Earth, we could detect the signal, which would appear as the short pulse that Bell had observed.

Antony Hewish was awarded (jointly with fellow astronomer Sir Martin Ryle) the 1974 Nobel Prize in physics for his work in identifying pulsars as a new class of stars. This was the first time the prize had been given for observational astronomy. Jocelyn Bell was ignored by the Swedish Nobel committee in spite of her stellar contributions. She earned her PhD in 1968. Pulsars appeared in the appendix of her dissertation.

Back to SETI. A completely new way of searching for an alien civilization based on the laws of physics—specifically, the second law of thermodynamics—was conjured by physicist and mathematician Freeman Dyson in 1960. At first, the core idea will seem like a grandiose science-fiction scheme (it was indeed inspired by science fiction), and its feasibility may be beyond our current engineering capabilities, but the underlying scientific logic is solid, and its technological scope is awe inspiring.

---

[3] Neutron stars are small but very dense objects formed after the collapse of stars bigger than the Sun. The typical radius of a neutron star is of the order of 10 kilometers (6.2 mi); but they can have masses of about twice that of the Sun and all that mass is packed into a region of size comparable to a small city. The pressure is so great that electrons and protons are squeezed together into neutrons. Thus, a neutron star is made up almost entirely of neutrons.

CHAPTER 2

The Dyson Sphere

• • •

> As the eons advanced, hundreds of thousands of worlds were constructed, all of this type, but gradually increasing in size and complexity. Many a star without natural planets came to be surrounded by concentric rings of artificial worlds. In some cases, the inner rings contained scores, the outer rings thousands of globes adapted to life at some particular distance from the Sun.
>
> —Olaf Stapledon

IN HIS 1960 ARTICLE IN the journal *Science* entitled "Search for Artificial Stellar Sources of Infrared Radiation" (Dyson 1960, 1667–68), Freeman Dyson—the English-born theoretical physicist and mathematician famous for his work in quantum electrodynamics, solid-state physics, astronomy, and nuclear engineering—argued that there may be another way to detect uncommunicative alien civilizations. The abstract read as follows: "If extraterrestrial intelligent beings exist and have reached a high level of technical development, one by-product of their energy metabolism is likely to be the large-scale conversion of starlight into far-infrared radiation. It is proposed that a search for sources of infrared radiation should accompany the recently initiated search for interstellar radio communications."

Dyson realized that advanced alien civilizations could be recognized by tracking their waste heat, detectable as infrared radiation (infrared excess). For example, see figure 2-1, where the infrared excess for Earth is shown as a result of human activity. Similarly, civilizations must, by the second law of thermodynamics, emit waste heat even if they are uncommunicative and try to conceal their existence. Dyson further argued that really advanced civilizations would re-engineer their solar systems, perhaps dismantling planets to form a shell of satellites around their star to capture its energy. In Dyson's own words, "One should expect that, within a few thousand years of its entering the stage of industrial development, any intelligent species should be found occupying an artificial biosphere which completely surrounds its parent star" (Dyson 1960, 1667–68).

Figure 2-1: Earth has 0.01 percent excess mid-infrared radiation because of human activity.

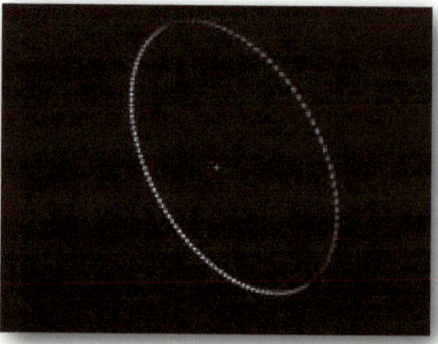

Figure 2-2: A Dyson ring—the simplest form of the Dyson swarm—to scale. Orbit is 1 AU in radius, collectors are 1.0 ×107 km in diameter (~25 × the Earth–Moon distance), spaced 3 degrees from center to center around the orbital circle. Image credit: www.wikipedia.com.

Dyson speculated that such an artificial biosphere would be the logical consequence of the long-term survival of our species. Not too far into the foreseeable future, all of Earth's nonrenewable resources, such as fossil and nuclear fuels, will be exhausted. Even the renewable sources will be unable to cope with growing demand. This is because Earth receives about one billionth of the total radiant output of the Sun, according to Berkeley astronomer Don Goldsmith, and humans utilize about one millionth of that. To sustain the future growth, the human race will need to capture much more of the Sun's light. It could do this with a Dyson sphere, a gigantic artificial structure, which he referred to initially as a shell, for the purpose of intercepting vast amounts of solar energy. If this could be done with great efficiency, the total energy gain could easily be many trillion times our current energy consumption on Earth.

The idea of an artificial biosphere was inspired by the science-fiction author Olaf Stapledon's *Star Maker* and by J. D. Bernal.

Many variants of the design of the Dyson sphere have been proposed; the designs of some are not even practical. Today, Dyson spheres are envisioned more as a dense network of orbiting solar-power satellites (collectors of solar energy) in the space around a star, capable of capturing most of or all the star's energy output. Dyson suggested that this method of energy collection would be inevitable for advanced civilizations whose goal is to ensure that a significant fraction of the star's energy impinges on a receiving surface of the collectors that could be used to the civilization's benefit. The simplest form of Dyson sphere—a ring of solar-power collectors at a distance from of, say, one hundred million miles from a star—is shown in figure 2-2. The central dot represents a star. This configuration is called a Dyson ring. With time a civilization might continue to add more Dyson rings to the space around its star, creating an incredibly powerful Dyson sphere (figure 2-3).

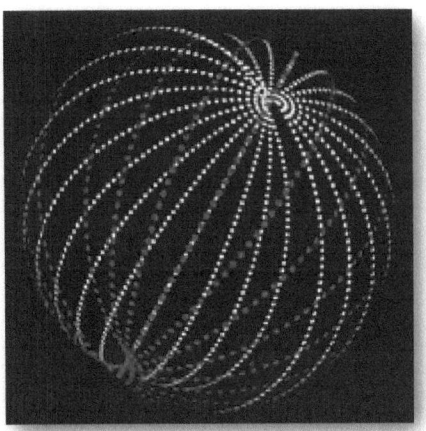

Figure 2-3: A relatively simple arrangement of multiple Dyson rings of the type pictured above, to form a more complex Dyson swarm. Rings' orbital radii are spaced 1.5 ×107 km with regard to one another, but average orbital radius is still 1 AU. Rings are rotated 15 degrees relative to one another, around a common axis of rotation. Credit: www.wikipedia.com.

Detecting Dyson spheres around other stars is challenging and might not be possible with today's technology. But that did not stop scientists from trying, and the literature contains several attempts to find the expected waste energy in mid-infrared wavelengths that should be emanating from a Dyson sphere. Searches by Vyacheslav Slysh at the Space Research Institute in Moscow in 1985 and by Richard Carrigan at Fermilab in 2009, using data from the Infrared Astronomical Satellite (IRAS), produced no results. Jugaku, Noguchi, and Nishimura (Jugaku, Noguchi and Nishimura 1995, 381–85) searched numerous stars with a 1.26-meter infrared telescope in Japan and examined IRAS data looking for a ten- to twenty-micron infrared excess. They did not find any candidates either. Null results were also reported by Jason Wright at Pennsylvania State University, University Park, and Matt Povich at the California State Polytechnic University in Pomona[4]. They used data from NASA's Wide-Field Infrared Survey Explorer (WISE) and the Spitzer Space Telescope, launched in 2009 and 2003 respectively, but found nothing.

With all the false starts, the question now is, how likely are we to ever find intelligent life somewhere in our galaxy, or in the universe for that matter?

To answer that question, one needs to be able to estimate the possible number of civilizations in our galaxy first and identify the areas to focus on in the search for alien life. That is done by an empirical equation called Drake equation, the subject of our next chapter.

---

[4] See (Gilster 2015).

CHAPTER 3

The Ranking of Civilizations

• • •

It isn't only the beauty of the night sky that thrills me. It's the sense I have that some of those points of light are the home stars of beings not so different from us, daily cares and all, who look across space with wonder, just as we do.

—Frank Drake

The Drake Equation

In 1961, Frank Drake conceived of an equation to estimate the number of civilizations, *N*, in our galaxy with which communication might be possible (Drake 1961, 40–46):

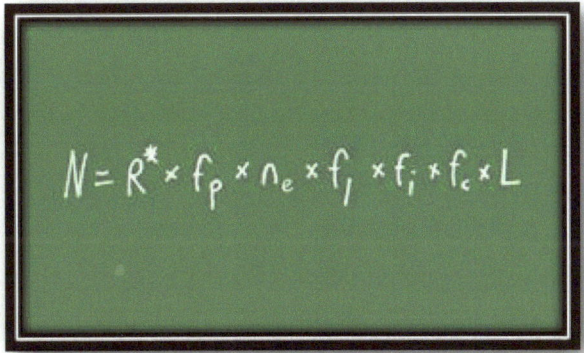

$$N = R^* \times f_p \times n_e \times f_l \times f_i \times f_c \times L$$

R^* = the average rate of star formation per year in our galaxy
f_p = the fraction of those stars that have planets
n_e = the average number of planets that can potentially support life per star that has planets
f_l = the fraction of the above that actually go on to develop life at some point
f_i = the fraction of the above that actually go on to develop intelligent life
f_c = the fraction of civilizations that develop a technology that releases detectable signs of their existence into space
L = the length of time such civilizations release detectable signals into space

The Drake equation is a seven-variable equation that has no unique solution. Nevertheless, it is a generally accepted tool for the scientific community to obtain a reasonable estimate for N, but one must make some drastic assumptions regarding the values of its variables.

The equation appears daunting at the first glance, but it is really a straightforward and clever approach to estimate the number of technological civilizations that may exist in our galaxy. The Drake equation's real worth is perhaps not in the answer it gives, but in the insights it provides, when one tries to plug in values of the various terms. The first three terms—R^*, f_p, and n_e—are fairly well known to astronomers, thanks to the recent advances in telescope and satellite technologies. For example, R^*, the average rate of star formation per year in our galaxy, is about one star per year, contrary to Drake's originally suspected ten in 1960, based on the data available at that time. Similarly, the estimates for f_p, or the fraction of those stars with planets, is pretty well known now: 0.4 or 40 percent. Finally, the current estimate for n_e, the average number of those planets that may develop an ecosystem, is between 0.5 and 2, contrary to Drake's original estimate of 2.

However, trying to estimate values for the last four terms of the equation gets trickier as lot more research and data are needed to provide reasonable answers. The fraction of those planets that succeed in developing life, f_l, is completely unknown. Is life inevitable, or is it a lucky anomaly? This is a hotly debated philosophical question as we shall see soon, but Drake believed that formation of life is inescapable under suitable conditions; he set f_l to 1.

Far more difficult to answer is the question of what fraction of those planets develop intelligent life, f_I. Drake put down 0.01, but this really is anybody's guess, and modern science can do no better. It is possible that Drake could be considerably off the mark here as some people argue, based on the fact that only one species evolved to be intelligent (possessing interstellar communication technologies) out of the millions found on Earth.

Also unknown is the value of the penultimate term, f_c, or the fraction of those planets with intelligent life that develop interstellar communication capabilities. Drake guessed 0.01, or 1 percent. But nobody really knows for sure until other alien civilizations are detected.

The last term, L, the average length of time such civilizations survive and continue to send communications, is perhaps the most interesting. The renowned British theoretical physicist Stephen Hawking believes that life on Earth is at a risk of being destroyed by a disaster, such as global climate change, an unexpected nuclear holocaust, a global pandemic, a genetically engineered virus, or even the growing threat of artificial intelligence. The risk is cumulative. Hawking, speaking at Oxford University Union, explained, "Although the chance of a disaster to planet Earth in a given year may be quite low, it adds up over time, and becomes a near certainty in the next thousand or ten thousand years. By that time, we should have spread out into space, and to other stars, so a disaster on Earth would not mean the end of the human race."[5] It is, therefore, not unreasonable to expect that an advanced civilization would face similar existential risks. In fact, the last term in the Drake equation, L, the average length of time such civilizations survive and continue to send communications, is very hard to guess as we have just one example to work with.

---

[5] Quoted in (Criss 2016).

Hawking's view of the future of life on Earth may sound apocalyptic. But the reality is that even if humankind survives all natural disasters, all self-induced catastrophes, and random celestial events such as impacts by comets or asteroids, life on Earth is not perpetual. It will eventually come to an end with the death of its Sun. In its dying phase in a few billion years, the Sun will become a red giant, absorbing its nearest planets Mercury, Venus, and Earth. But Earth will become uninhabitable much sooner than that. Its oceans will evaporate, making the planet a scorched, lifeless wasteland before being finally engulfed by the Sun.

Nevertheless, using his own equation, Frank Drake's current estimate for communicative civilizations in our Milky Way galaxy comes out to be around ten thousand. But your guess may be as good as his! In fact, you can use Drake's equation to make your own estimate by visiting the PBS website (Ferris n.d.). Scientists now have a better understanding of some of the variables in his equation, and they have been continually revising his numbers. A conservative estimate today is that there are between two and fifty thousand. More than N, the Drake equation is also extremely useful for identifying the areas one must focus on in the search for alien life.

The Drake Equation Revisited

More precise estimates of some of the factors in the Drake equation were attempted, and several improvements to the equation were proposed over the years, particularly after the discovery of exoplanets. We will discuss exoplanets later in this book. Here I will mention one recent approach by Frank and Sullivan (2016, 359–62), who showed how to completely eliminate the least known term L—how long civilizations might survive—by simply expanding the equation. This allowed them to answer the cosmic archeological question, "Have they ever existed?" rather than the answer that the Drake equation gives to the usual question of narrower scope, "Do they exist now?"

That still did not get rid of the huge uncertainties in f_i, the fraction of those habitable planets where advanced life could evolve. Rather than trying to guess the value of f_i, Frank and Sullivan proceeded to calculate the odds against it. As a result, a lower limit on the probability that technological species have ever evolved anywhere other than on Earth emerged. How did they achieve this? In their own words, they were able to segregate the "newly measured astrophysical factors from the fully unconstrained biotechnical ones" (Frank and Sullivan 2016, 359–62) in the Drake equation, with the help of the new data obtained from NASA's Kepler satellite and other searches.

The cosmic archeological question is of particular scientific and philosophical consequence according to Frank and Sullivan. In their own words, "We find that as long as the probability that a habitable zone[6] planet develops a technological species is larger than $\sim 10^{-24}$, then humanity is not the only time technological intelligence has evolved." In other words, any value lower than that would imply "we are likely alone and singular in the history of the observable universe." This leads to the definition of a pessimism line: a constraint that has important scientific and philosophical consequences. If the evolutionary processes lead to values higher than this lower limit, according to Frank and Sullivan, we are not the only instance in which the universe has hosted a technological species.

The Fermi Paradox

Where is everybody?

—Enrico Fermi

The above quote is attributed to the famous Italian-born American scientist Enrico Fermi, one of the most influential physicists of the twentieth century who, among other landmark achievements, directed the first controlled chain reaction involving nuclear fission.

---

[6] A habitable zone is the region surrounding a host star in which water can remain in its liquid state and planets could support life. See chapter 6 for more details.

If any among these fifty thousand civilizations produce cultures that are capable of colonizing over interstellar distances, even at a small fraction of the speed of light, the Milky Way galaxy should have been completely colonized in no more than a few million years. Since our galaxy is billions of years old, Earth should have been visited and colonized long ago. In fact, they should already have completed the expansion to fill the galaxy before the emergence of life from the ocean. The absence of any evidence of such visits is the Fermi paradox.

Interstellar distances are massive, perhaps too vast to be conquered by living beings whose lifetimes are only finite. A technologically advanced civilization would therefore be faced with the challenge of finding the most efficient way to travel interstellar or interplanetary distances; however, it should be able to construct self-reproducing, autonomous robots to colonize the galaxy. Mathematically, the most efficient way of exploring the hundreds of billions of stars in the galaxy is via the von Neumann probe. The Hungarian-born American mathematician John von Neumann was the first to develop a mathematical theory of machines that could make exact copies of themselves. Famed futurist and physicist Michio Kaku describes a von Neumann probe in his blog in the following way:

> A von Neumann probe is a robot designed to reach distant star systems and create factories which will reproduce copies themselves by the thousands. A dead moon rather than a planet makes the ideal destination for von Neumann probes, since they can easily land and take off from these moons, and also because these moons have no erosion. These probes would live off the land, using naturally occurring deposits of iron, nickel, etc. to create the raw ingredients to build a robot factory. They would create thousands of copies of themselves, which would then scatter and search for other star systems.
>
> Similar to a virus colonizing a body many times its size, eventually there would be a sphere of trillions of von Neumann probes

expanding in all directions, increasing at a fraction of the speed of light. In this fashion, even a Galaxy 100,000 light years across may be completely analyzed within, say, a half million years. (Kaku n.d.)

Back in the 1940s, Fermi realized that the aliens should have had more than enough time to colonize the galaxy. The apparent absence of evidence of extraterrestrials in the vicinity of Earth is thus clearly a paradox. SETI has been going for nearly fifty years, employing increasingly powerful telescopes, faster computers, and data-mining techniques. But the reality is that it has so far consistently come up with null results. Where is everybody?

There are several possible solutions to Fermi's paradox. Physicist Brian Cox has explained some of them in his article "Were We Contacted by Aliens in 1977?" on BBC online:

> **Space is too vast.** We can't survey every part of the night sky, and it's possible that radio signals will become too weak as they cross our galaxy to be detected.
>
> **Civilizations don't last.** Advanced civilizations may tend to destroy themselves, through war or environmental catastrophe. If not, a disaster like an asteroid strike might wipe them out.
>
> **Our planet is unique.** According to the rare Earth theory, even if simple life has arisen elsewhere, the chance events that led to intelligent life on our planet is highly improbable (Cox 2017, 7).

Clearly, the reason for the lack of evidence of extraterrestrial civilization—or the Great Silence, as it is sometimes called—falls under the realm of abstract speculation. The Fermi paradox will be revisited later in this book (see chapter 5), but first we need to be able to rank the detectable civilizations as guessed by Drake's equation, an important working tool in the field of SETI, to better understand what distinguishing features they might possess. That will tell us how to properly set the search criteria.

Kardashev Civilizations

> The material factors which ultimately limit the
> expansion of a technically advanced species are the
> supply of matter and the supply of energy.
>
> —Freeman Dyson

A scheme for classifying advanced technological civilizations was proposed by the Russian radio astronomer and SETI theorist Nikolai Kardashev. His original classification of the three types of civilizations, based on their energy consumption, is given below:

I —technological level close to the level presently attained on Earth, with energy consumption at ~ 4×10^{19} erg/sec.

II —a civilization capable of harnessing the energy radiated by its own star (for example, the stage of successful construction of a Dyson sphere); energy consumption at ~ 4×10^{33} erg/sec.

III —a civilization in possession of energy on the scale of its own galaxy, with energy consumption at ~ 4×10^{44} erg/sec. (Kardashev 1964, 217–21).

Kardashev calculated how powerful an extraterrestrial radio signal would have to be in order to be detected by conventional radio astronomical techniques. The numbers he came up with were rather large and formed the basis of his ranking of the civilizations.

Since the publication of Kardashev's seminal paper, there have been various extrapolations of Kardashev's original classification by scientists. The following quote is from David Darling's website, the Encyclopedia of Astrobiology, Astronomy, and Spaceflight:

> A Type I civilization would be able to marshal energy resources for communications at a planet-wide scale, equivalent to the

entire present power consumption of the human race, or about 10^{16} watts. A Type II civilization would surpass this by a factor of approximately ten billion, making available 10^{26} watts, by exploiting the total energy output of its central star. Freeman Dyson, for example, has shown in general terms how this might be done with a Dyson sphere. Finally, a Type III civilization would have evolved far enough to tap the energy resources of an entire Galaxy. This would give a further increase by at least a factor of 10 billion to about 10^{36} watts. [Carl] Sagan pointed out that the energy gaps between Kardashev's three types were so enormous that a finer gradation was needed to make the scheme more useful. A Type 1.1 civilization, for example, would be able to expand a maximum of 10^{17} watts on communications, a Type 2.3 could utilize 10^{29} watts, and so on. He estimated that, on this more discriminating scale, the human race would presently qualify as roughly a Type 0.7. (Darling, Kardashev civilizations n.d.)

Human civilization is not even in Kardashev's scale yet.

Currently, human civilization is able to harness only a portion of the energy that is available on Earth. Although intermediate values were not discussed in Kardashev's original proposal, Carl Sagan calculated humanity's civilization type to be 0.7 in 1973 by interpolating and extrapolating the values. The current state of human civilization has been named Type 0.

In the table below, the International Energy Agency's past and projected values for planetary power production yield are shown against the corresponding Kardashev scale estimates.

Year	1900	1970	1973	1985	1989	1993	1995	2000	2001	2002	2004	2010	2030
Terawatts	0.67	0.6	8.2	9.2	10	11	12	13	13	14	14	16	22
Kardashev Scale	0.58	0.68	0.69	0.7	0.7	0.7	0.71	0.71	0.71	0.71	0.72	0.72	0.73

If the data is fit to a linear model, Type 1 status in Kardashev's scale will be reached around the year 2250, when human civilization is expected to harness most forms of energy available on Earth. This is consistent with Freeman Dyson's estimates that, within two hundred years or so, we should attain Type I status.

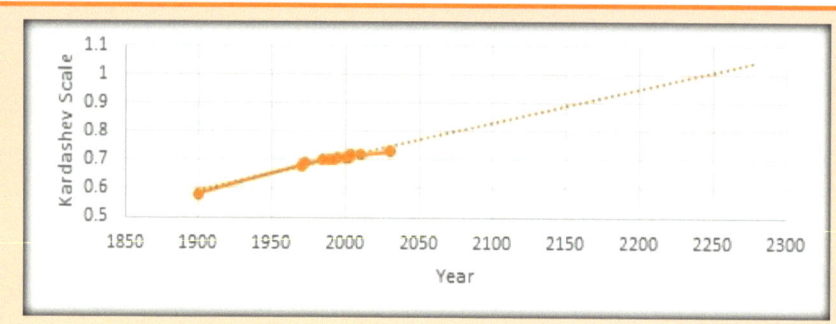

In fact, growing at a modest rate of 1 percent per year, Kardashev estimated that it would take 3,200 years to reach Type II status and 5,800 years to reach Type III status.

According to Kardashev, the possibility of detecting a Type I civilization using an Ozma-like search would be extremely low. Instead, he suggested SETI programs should concentrate on looking for the kind of intense radio signals that might originate from Type II or Type III activity. Of particular interest were two radio sources whose California Institute of Technology designation numbers were CTA-21 and CTA-102.

In 1964, G. B. Sholomitskii[7] used the Crimea Deep Space Station to examine the radio source CTA-102 at a frequency of 923 megahertz and reported periodic variations in its radio flux over a period of 102 days in accordance with Kardashev's claim that this was indicative of a possible signal from a Type III civilization; however, subsequent studies found the variability of radio flux of CTA-102 had another explanation. CTA-102 was actually a quasar, and further studies led to the discovery of a new and fundamental property of quasars—the variability of their radio emissions.

---

[7] See (Braude, et al. 2012, 97).

CHAPTER 4

The Great SETI Debate

• • •

When I heard the learn'd astronomer,
When the proofs, the figures, were ranged in columns before me,
When I was shown the charts and diagrams,
to add, divide, and measure them,
When I sitting heard the astronomer where he lectured
with much applause in the lecture-room,
How soon unaccountable I became tired and sick,
Till rising and gliding out I wander'd off by myself,
In the mystical moist night-air, and from time to time,
Look'd up in perfect silence at the stars.

—WALT WHITMAN

DO GALACTIC CIVILIZATIONS EXIST?
There are two categories of thinkers:

1. **Contact optimists.** Those who contend that simple reasoning indicates a universe in which life and intelligence are commonplace.
2. **Proponents of uniqueness hypothesis.** Those who suggest that Earth is probably the first and only abode for technical civilization in our galaxy.

A philosophical debate has always raged between these two categories of thinkers. Physicist Frank Tipler, a proponent of the uniqueness hypothesis, assumed the von Neumann probe would be commonly adopted by advanced civilizations (if they exist) for galactic colonization because it is the fastest and most economical means of doing so. He proposed a conservative value of three hundred million years, or less than 5 percent of the present age of the galaxy. Thus, there should be a significant and obvious presence of such devices within the solar system. Yet no such presence has been detected. Tipler, therefore, concludes that we are the only intelligent society among the galaxy's several hundred billion stars.

The validity of this conclusion has been disputed by, among others, planetary scientists Carl Sagan and William I. Newman. Their objection is based on the cosmological principle: our position in the universe cannot be preferred in any sense. It follows from the cosmological principle that what was possible on Earth should be possible elsewhere. This is the problem of the locality or nonlocality of the existence of life. The locality of life would mean that life is not common phenomenon.

Sagan and Newman's objection is a largely philosophical. Their arguments were succinctly summarized by David Darling in his website:

> [Sagan and Newman] first pointed out that Tipler had underestimated the number of von Neumann probes there ought to be. With exponential growth, a single self-replicating probe could be expected to convert the entire mass of the Galaxy into copies of itself within 2 million years. Any species intelligent enough to build such a probe, Sagan and Newman argued, would also be intelligent enough to realize the danger of it and so would not embark upon the project in the first place. In the event of a von Neumann probe being released either accidentally or maliciously, it would be a prime duty of other, responsible civilizations, said Sagan and Newman, to stamp out the "infection" before it could spread. (Darling, The Worlds of David Darling n.d.)

The two scientists also argued that intelligent aliens might refrain from constructing fleets of self-replicating von Neumann probes because such machines might "cannibalize" their creators. Advanced aliens might have "much more exciting and fulfilling objectives…than strip-mining or colonizing every planet in sight." Only benevolent civilizations are "pre-adapted to live with other groups in mutual respect." And our Milky Way is packed with advanced civilizations: "We think it is possible that the Milky Way is teeming with civilizations that are far beyond our level of advance as we are beyond the ants; and paying us about as much attention as we pay to the ants" (Sagan and Newman 1983, 113–21).

A similar argument was put forth by Michio Kaku: "Let's say we have an ant hill in the middle of the forest. And right next to the ant hill, they're building a ten-lane super-highway. And the question is: Would the ants be able to understand what a ten-lane super-highway is? Would the ants be able to understand the technology and the intentions of the beings building the highway next to them?"[8] In other words, higher civilizations may exist all around us, but our current technology reach is too primitive to perceive them.

On the other hand, the rare Earth hypothesis is commonly used as an argument for the apparent absence of inhabited planets. According to this hypothesis, evolution of life and, eventually, of intelligent life on Earth are due to a large number of fortuitous events, which include the following:

- The age, size, and composition of our Sun.
- The location of Earth and the inclination of its axis.
- Earth's powerful magnetic field, caused by its nickel-iron core coupled with its rapid rotation, protects it from the effects of solar wind.
- Abundance of water and a stable oxygen-rich atmosphere.
- Development of a carbon-based chemistry requires a fine adjustment of physical constants, such as the electric charge and even the dimension of space-time. If these constants had significantly

---

[8] Quoted in (Urban 2014).

different values, either the nucleus of the carbon atom would not be stable or the electrons would collapse into the nucleus.
- Earth's active interior. Movement of the tectonics plates was responsible for creation of the major mountain ranges such as the Alps, the Himalayas, and the Andes, giving rise to different ecological conditions favorable for the propagation of a great variety of species.
- The presence of the Moon, Jupiter, and Saturn as shields for the bombardment of comets and meteorites during the early stages of Earth.
- Long ice ages and numerous and fortuitous catastrophes, such as the one sixty-five million years ago, which led to the extermination of dinosaurs, paving the way for more diversified and complex life-forms.

Proponents of the rare Earth hypothesis, such as Peter Ward and Donald Brownlee of the University of Washington, believe that complex life is extremely rare in the universe. In Ward and Brownlee's book *Rare Earth* (2000), the scientists argue that the essential ingredients for life are a terrestrial planet with tectonic plates located in the habitable zone of a suitable star, a large moon, magnetic field, and oxygen. A massive planet like Jupiter is also necessary as a gatekeeper to detect killer asteroids. Microbial life may be common in the universe, but for complex intelligent life to develop and prosper, an extraordinary set of unlikely circumstances is crucially needed. Life on Earth evolved unusually fast and can be regarded as a cosmic aberration. The evidence gathered from other planets in our solar system suggests that the chances of life developing on an average Earthlike rocky world are exceedingly low.

The prokaryotic bacteria, the first primitive life-forms on Earth, appeared about five hundred million years after the cooling off of Earth's crust and once the epoch of dense bombardment of meteorites and comets ended. These bacteria were the only life-forms during the first 2 billion years of Earth's 4.6-billion-year history. Mammals, including apes and humans, appeared much later, only after the extinction of the dinosaurs sixty-five million years ago. Proconsul, an extinct genus of primates with whom gibbons, great apes, and humans share a common ancestral lineage, existed from twenty-three to twenty-five million years ago during the Miocene epoch. The *Australopithecus*, one of the longest-lived and best-known early human species, existed only three to four million years ago. In short, it took almost four billion years, or about 96 percent of the age of Earth, for intelligent life to evolve, which is a long time, even on a cosmic scale.

Distinguished evolutionary biologist Ernst Mayr highlighted the enormous complexity of DNA and RNA and their functions in the production of proteins, the basic building blocks of life. He estimated that the likelihood of similar biological developments happening elsewhere in the universe is almost nil.

Needless to say, proponents of both schools of thought have attempted to solve the Fermi paradox in their favor, using highly speculative arguments. With so little experimental evidence, it is virtually impossible to ascertain which viewpoint is correct. It depends on personal biases and other considerations. Sagan and Newman acknowledged that as well. Their paper "The Solipsist Approach to Extraterrestrial Intelligence" ends with the following lines:

> The question touches on religious and political matters where predispositions have traditionally played important roles. But it is abundantly clear from the history of science that no convincing resolution of this issue is likely to come from protracted debates carried on with great passion and sparse data. We have an alternative denied to the medieval scholastics: we are able to experiment. We can organize a scientifically rigorous systematic search for extraterrestrial intelligence using the technology of modern radio-astronomy. That is where the energies should be focused of those concerned with the great issue of the existence of other technical civilizations in the cosmos. (Sagan and Newman 1983, 113–21)

Sagan's and Newman's dream of a "scientifically rigorous systematic search" has recently become a reality but using a completely unrelated methodology. This search, not as much to find extraterrestrial civilizations as to find signs of life in our galaxy and beyond, is the topic of our next two chapters.

CHAPTER 5

The Search for Alien Life in Our Solar System

• • •

I think we're going to have strong indications of life beyond Earth within a decade, and I think we're going to have definitive evidence within 20 to 30 years.

—Ellen Stofan

WATER IS THE ESSENTIAL INGREDIENT of life as we know it.

Water, or H_2O, is abundant in our solar system. Oceanic icy worlds have been found in Jovian moons Europa, Callisto, and Ganymede; in Saturn's moons Enceladus and Titan; and probably in Neptune's moons Triton. Possible past ocean worlds that are now extinct or are remnants of subsurface oceans were found in the asteroid Ceres; in the icy satellites Ariel, Iapetus, Umbriel, Titania, Oberon, and Triton; and in large dwarf planets such as Pluto, Charon, Eris, Sedna, and Orcus.

Some of these watery worlds may contain the essential ingredients for life—the fluxes of chemical energy, water, and raw ingredients that make up cellular material and drive biogeochemistry. The conventional thinking is that if you want to find evidence of alien life, follow the trail of alien water. In spite of the immense technological challenges of sending probes into these alien worlds, a growing number of scientists now agree that we must try.

The caveat is that the alien oceans have to be studied using methods developed to study the movement of energy and nutrients in Earth's own systems. Nevertheless, if evidence of life is ultimately found lurking somewhere in these icy moons where the temperatures hover close to absolute zero, it would most definitely be a whole new class of habitability in our solar system and probably the most common form in the universe.

In this chapter, we digress in our quest for extraterrestrial intelligence and concentrate on the more mundane but equally important question, "Is there any alien life-form beyond Earth in our own solar system albeit not the intelligent kind?" Jupiter's moon Europa is a good place to start looking, as it is the subject of intense scientific interest because of a large body of evidence that points to the existence of a vast, salty ocean underneath its icy surface that holds more than twice the amount of water of all the Earthly oceans combined.

Enigmatic Europa

Europa is one of the four moons of Jupiter discovered by Galileo Galilee on January 8, 1610 (Io, Ganymede, and Callisto are the other three). It is the smallest of the four Galilean moons, slightly smaller than Earth's own Moon. Nevertheless, Europa is one of the more intriguing satellites in our solar system because of the strong belief that it has a hidden ocean that is in contact with its rocky seafloor.

Figure 5-1: Europa's stunning surface. Image credit: NASA.

Europa came into the spotlight in 1979, when NASA's *Voyager 2* mission flew past it, sending stunning images of its surface, which is smooth and bright, consisting of water and ice crisscrossed by long, linear fractures. While the indication of this internal ocean is quite strong, its presence still awaits confirmation. This Jovian moon could therefore be a promising place to look for life beyond Earth if this ocean is proven to exist on a future mission.

Europa's orbital period around Jupiter is 3.5 days. It is tidally locked with Jupiter, which means that the same hemisphere of the moon always faces the planet. Being in an elliptical orbit, Europa's distance from Jupiter varies. This causes an uneven gravitational pull between the near and far sides of the moon and triggers tides that stretch and relax its surface. The magnitude of this difference changes as Europa orbits around Jupiter. The linear fractures across Europa's surface are caused by the tides, which also supply energy to the moon's icy shell. If Europa's ocean exists, the tides might also create volcanic or hydrothermal activity on the seafloor, supplying nutrients that could make the ocean suitable for living things.

Jupiter and its mysterious moons were explored by NASA's Galileo mission between 1995 and 2003. During the eleven flybys of Europa, the satellite obtained much closer images of the moon's fractured surface (figure 5-1), revealing hard layers of ice interlaced with glacial ridges and domes that carved out paths across an icy surface concealing a deep ocean of liquid water. The highly disrupted areas are called chaos terrains, where the landscape is covered with a mysterious reddish material consisting of broken blocks of ice that drifted and refroze on the surface.

Adding to Europa's mysticism was something that the researchers using the Hubble telescope saw in 2012—a huge vapor cloud hovering above the moon's southern hemisphere (figure 5-2). More water plumes were found in 2016; their intensity varied according to where Europa was located in its orbit. Active jets were spotted when the moon was farthest from Jupiter, and none when it was closer. It appeared that the water vents narrowed or closed when the moon was very close to the gas giant because of its massive gravitational pull. They opened up when the moon was farthest from the gas giant.

Discovery of the water plumes was a significant result, as it showed that liquid water is able to break through the crust and spray far and wide into space. Either liquid water is close to the surface, or there are very deep raptures in the icy crust. The reddish-brown markings on Europa's surface could be the deposits of oceanic salts bleached by the intense radiation arriving from Jupiter. But the mineral-rich liquid water is particularly suitable for life to prosper inside the deep underground ocean.

Figure 5-2: Water vapor detected over Europa's south pole in observations taken by NASA's Hubble Space Telescope in December 2012. Credits: NASA/ESA/L. Roth/ SWRI/University of Cologne.

NASA is already developing *Europa Clipper*, a spacecraft intended to launch in the 2020s to perform forty-five flybys of Jupiter's moon to investigate in depth whether the icy moon might be habitable someday. Exploring Europa poses a huge challenge because it is immersed in Jupiter's giant radiation belts. The plan is to place the radiation-tolerant *Europa Clipper* in a long, looping orbit around Jupiter to perform the repeated close flybys. Alternatively, the spacecraft could be placed in an orbit around Europa itself, but the problem is the constant barrage of damaging radiations from Jupiter. By abandoning the idea of orbiting Europa, the probe will avoid the devastating radiation by flying past Europa multiple times, coming as close as fifteen miles above its frigid surface, and then retreating to a distant part of the orbit where radiation levels are considerably lower. Nine science instruments have been selected to investigate whether the mysterious icy moon could harbor conditions suitable for life. During these flybys, the mission will take high-resolution photographs of Europa's surface using an onboard imaging system; determine the composition of the icy moon's surface with a penetrating radar; study its faint atmosphere with a mass spectrometer; and investigate its ice shell, ocean, and interior. The water plumes have provided scientists a great opportunity to sample Europa's store of liquid water to look for chemical signatures of microbial life without drilling holes into its icy surface, which is technologically challenging and considerably more expensive; humans have never landed a spacecraft on an ice world before. Instead, *Europa Clipper* will simply glide through the plumes and measure its composition using the equipment on board.

THE OCEAN WORLD OF ENCELADUS

Saturn's moon Enceladus is the brightest world in our solar system and a source of wonder for astronomers and planetary scientists. As recently as 2015, NASA's Cassini probe revealed that Enceladus is an active, tiny moon with an ice-covered ocean beneath its crust, similar to Jupiter's moon Europa.

Cassini's transmitted images show a plume of material escaping into space at about eight hundred miles per hour through cracks in its icy shell (figure 5-3). The plume extends hundreds of miles into space; some of the material drops back onto Enceladus while some escapes to form Saturn's vast E ring. In October 2015, the spacecraft flew directly into the plume of escaping material and sampled its chemical composition using

Figure 5-3: Cassini images of Saturn's moon Enceladus backlit by the Sun show the fountain-like sources of the fine spray of material that towers over the south polar region. Credits: NASA.

the onboard ion-neutral mass spectrometer. Molecular hydrogen native to Enceladus—a potential foodstuff for bacteria and a sign of hydrothermal activity—was detected in the plume. Hydrogen is released in hydrothermal reactions between the minerals inside rocks and organic materials.

The relatively high hydrogen abundance in the plume is a sign that the water in Enceladus's ocean reacts with its rocks via hydrothermal processes in vents deep beneath Enceladus's icy shell. This drives the ocean out of chemical equilibrium, just like the water inside Earth's hydrothermal vents, potentially providing a source of chemical energy.

The presence of hydrothermal vents in Enceladus can also be inferred from the evidence obtained by analyzing the composition of Saturn's E ring, which is mostly composed of droplets of ice. Cassini found silica nanoparticles, about ten nanometers in size, interspersed with the ice droplets, as well as traces of methane, ammonia, carbon monoxide, carbon dioxide, simple organic compounds, and salts. The silica nanograins were linked to Enceladus, where water–rock interactions occur at temperatures above 90°C (194°F), pointing to the presence of these hydrothermal vents beneath Enceladus's icy shell, similar to the hydrothermal vents found in the ocean floor here on Earth.

Why is the water so warm in such icy surroundings? The combined gravitational attraction of Saturn and the other moons makes Enceladus wobble in its path, causing it to heat up. But current models show that gravity alone cannot explain the warm oceans; there must be something else that generates the heat. The obvious culprit is the ongoing chemical reactions between the rocks inside Enceladus and the water, leading to the speculation that the generated heat could possibly create the conditions that can be exploited by microbial life, just the way primitive organisms on Earth live off energy from the deep ocean vents where sunlight can never reach. Microorganisms use a process called methanogenesis—the biological production of methane mediated by these microorganisms, commonly called methanogens. The production of methane is the energy-yielding metabolism of methanogens that does not require sunlight and is unique to these organisms. In fact, a whole ecosystem can exist supported by methanogens, rather than by the photosynthetic plants we are so familiar with.

On the other hand, the large amount of molecular hydrogen detected could also make life less likely in Enceladus. Mary Voytek, senior scientist of NASA astrobiology, explains: "[The] fact that that we can measure such high concentrations of hydrogen and carbon dioxide mean that there might not be life there at all, and if there is life, it's not very active…We have this buildup of food that's not being used. And part of that could be that we think Enceladus might be fairly young."[9]

On the other hand, the Jovian moon Europa is billions of years older than Enceladus—it may be a more favorable place for life to spawn, as the same process seen in Enceladus could be in play in Europa as well, according to Voytek. "My money for the moment is still on Europa," she says.

Needless to say, all present speculations are farfetched, and it will take scientists years to definitively establish whether life indeed exists on either moon. More clues could emerge later this year (2017) when the Cassini spacecraft flies past Enceladus at a distance of just thirty miles before plunging into Saturn on its final pass. Then in 2018, the James Webb Space Telescope (JWST) is scheduled to be launched, and one of its early goals is to study the plumes of Europa and Enceladus. JWST is optimized to measure infrared light. It should, therefore, be able to determine the chemical composition of the plumes, as certain chemical imbalances in its spectroscopic data could be indicative of a habitable environment or even of alien life. However, definitive evidence will have to wait till the data from the *Europa Clipper* mission, set to launch in the 2020s, is available.

ORGANIC MATERIAL ON CERES

Some fundamental questions that science needs to answer are, "How did life originate on Earth? Where did some of the organic molecules that are the building blocks of life come from?" The current theory is that Earth was formed in the protoplanetary disk, where the temperature of the mixture of gas and dust was too high for water vapor and some more volatile organic components to condense. This has led to the idea that those substances may have been delivered to Earth by asteroids and comets from the outer solar system.

---

[9] Quoted in (Kennedy 2017).

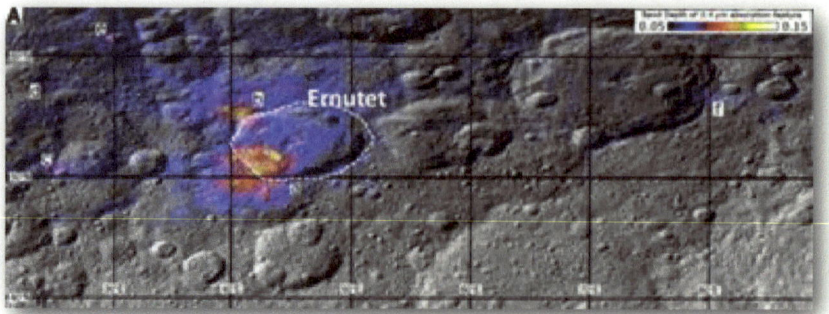

Figure 5-4: Dawn spacecraft data show a region around the Ernutet crater where organic concentrations have been discovered (labeled a through f). The color coding shows the strength of the organics absorption band, with warmer colors indicating the highest concentrations. Image courtesy of NASA/JPL-Caltech/UCLA/ASI/INAF/MPS/DLR/IDA.

Recently, NASA's *Dawn* spacecraft has detected organic compounds on Ceres, a dwarf planet and the largest object located in the main asteroid belt between Mars and Jupiter. Computer models predict that a substantial amount of water ice is present in its mantle and outer shell. The Herschel telescope and NASA's *Dawn* spacecraft have observed the release of water vapor from Ceres. The exposed water and ice is mainly localized on a broad region of approximately one thousand square kilometers close to the approximately fifty-kilometer Ernutet crater (figure 5-4). The combined presence on Ceres of ammonia-bearing hydrated minerals, water ice, carbonates, salts, and organic material indicates a very complex chemical environment, suggesting environments favorable to prebiotic chemistry. After examining the infrared spectra taken by the *Dawn* spacecraft as it orbited Ceres, an international team of researchers has ruled out an external origin, such as an impact, suggesting that the material must have formed on Ceres itself. As to how the molecules could have formed and migrated to the surface, that is a topic for further research.

The European Space Agency's *Rosetta* spacecraft has studied the comet 67P/Churyumov-Gerasimenko in detail since August 2014. Once again, complex organic molecules and even amino acids were found, leading to the conclusion that such molecules are ubiquitous on small bodies in the solar system and that water ice is abundant in the asteroid belt.

What Does All This Mean?

We conclude this chapter by noting that water is plentiful in our solar system. Water has been found in the many places, including on distant moons of the giant planets, asteroids, and some dwarf planets in the asteroid belt. NASA satellite missions have discovered liquid water underneath the frozen surfaces of two of Jupiter's moons, Europa and Ganymede, as well as in Saturn's moon Enceladus. Oceans covered much of Mars in the ancient past, and seasonal dark streaks observed on the Red Planet's surface today may be caused by salty flowing water. Since life as we know it needs water, these discoveries are significant and have raised the hope that some of these bodies may be harboring the life that humankind has long been searching for. If evidence of life is indeed found somewhere in our solar system, it will most likely be a whole new class of habitable environment bearing little resemblance to life on Earth.

NASA's *Dawn* spacecraft recently discovered that the chemistry needed to make life's building blocks is widespread in our solar system. Aliphatic organic molecules were spotted across a large swath of the dwarf planet Ceres (De Sanctis et al. 2017, 719–22). Scientists think that the compounds were likely produced on Ceres itself. This is a significant discovery, suggesting that the rich chemistry needed to create life's building blocks exists throughout the solar system.

It is important to point out, however, that no evidence of such life has actually been found anywhere except on Earth—all we have at the moment is some circumstantial evidence of the possibility of microbial life-forms. There still might be nothing at all. Life on Earth may be a singularity after all.

CHAPTER 6

Exoplanets

• • •

> We stand on a great threshold in the human history of space exploration. On the one side of this threshold, we know with certainty that planets orbiting stars other than the Sun exist and are common…If life is prevalent in our neighborhood of the Galaxy, it is within our resources and technological reach to be the first generation in human history to finally cross this threshold, and to learn if there is life of any kind beyond Earth.
>
> —Sara Seager

> But one day, many of us will gaze at the encyclopedia containing the coordinates of perhaps hundreds of earthlike planets in our sector of the galaxy. Then we will wonder, as Sagan did, what a civilization a million years ahead of ours will look like.
>
> —Michio Kaku

If our Sun has planets, shouldn't other stars have planets as well? Moreover, if one of the planets in our solar system harbors life, can't we expect to find life on some distant planet that revolves around a different star too? Many scientists and philosophers had intuitively assumed the existence of planets beyond our solar system. But it took a while to identify them.

The interest in discovering life-forms elsewhere in our galaxy, and in the universe, has received a huge boost recently with NASA's newfound ability to detect planets in solar systems other than ours. Twenty-five years ago, nothing was known about any other worlds outside our solar system. Thanks mainly to NASA's Kepler mission, which was launched on March 6, 2009, to explore our galaxy's interplanetary systems, we now know the Milky Way is teeming with planets, some of them resembling our own Earth.

A planet that does not orbit our Sun but orbits a different star or a brown dwarf is called an exoplanet. According to Wikipedia, more than 1,800 exoplanets have been discovered (1,885 planets in 1,184 planetary systems, including 477 multiple planetary systems) as of January 27, 2015. Almost all of these reside inside our Milky Way. But a few of them could be of extragalactic origin.

Extrasolar planets are indirectly detected by observing the effects that they have on the parent star.

The first extrasolar planet was discovered in 1994 by Dr. Alexander Wolszczan, a radio astronomer at Pennsylvania State University. Wolszczan had discovered two or three planet-size objects orbiting a pulsar in the Virgo constellation. He observed regular variations in the pulsar's rapidly pulsated radio signal that indicated the planets' complex gravitational effects on the dead star. It was quickly realized that these worlds were barren and inhospitable, as they are permanently bathed in high-energy radiation from the pulsar.

In 1995, two Swiss astronomers, Michel Mayor and Didier Queloz of Geneva, found a rapidly orbiting planet located close to 51 Pegasia, a star similar to our Sun. The planet was observed indirectly using the radial velocity method. The mass was determined to be between half and two times the mass of Jupiter.

These announcements marked the beginning of a flood of discoveries. Three months later, a team led by Geoffrey W. Marcy and Paul Butler of San Francisco State University and the University of California at Berkeley respectively, confirmed the discovery and added two more planets to the one discovered by the Swiss team. By the end of the twentieth century, several dozen exoplanets had been discovered by carefully observing nearby stars for several months or even years.

Today, we know that most of the exoplanets found so far are enormous gas giants such as Jupiter or Saturn in our own solar system. But most mind boggling is their variety. According to well-known planetary scientist and astrobiologist Sara Seager: "Some stars have a giant planet like Jupiter where the Earth would be. Other stars have planets like Jupiter 10 times closer to them than Mercury is to our Sun. Some stars have planets we call 'super-Earths,' rocky worlds bigger than Earth but smaller than Neptune" (Seager 2009, 5). A handful of them are believed to be Earth-size planets, and their numbers are growing rapidly.

The obvious question now is, "What are the chances these exoplanets could harbor life?" Our current understanding of life's origin on Earth indicates that given a suitable environment and sufficient time, life will develop on other planets.

The massive gaseous stars are not suitable for life because, in general, they have no surface and are too hot. One of the main requirements for life as we know it is liquid water. Water can remain in a liquid state between 273°K and 373°K, unless

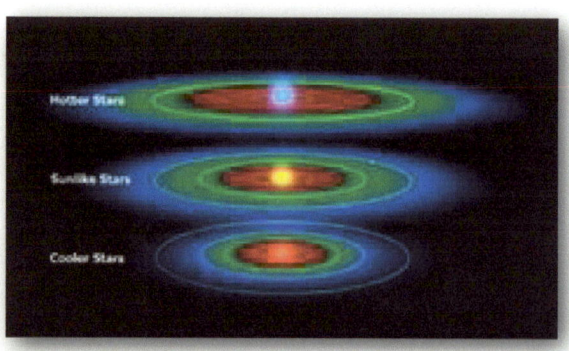

Figure 6-1: Habitable zone relative to size of star. Image credit: Wikipedia.

the pressure is too low, in which case the water turns into vapor. The region in a planetary system where the temperature is in this range is called the habitable zone or the Goldilocks zone. See figure 6-1 for a visualization of a habitable zone relative to a star. The red region is too warm, the blue region too cold, and the green region is just right for liquid water. However, there is another important consideration: most stars that are being searched for life-bearing planets must survive long enough for their planets to develop life.

The discovery of habitable exoplanets is of particular interest, especially Earthlike rocky worlds, which are the best candidates to start looking at for signs of life. To be Earthlike, a planet must be both Earth-size (less than 1.25 times Earth's girth and less than twice Earth's mass) and must circle its host star inside the habitable zone. Table 1 shows a list of potentially habitable Earthlike planets identified as of August 2015, branded based on the above criteria. Public interest piques every time the discovery of an Earthlike exoplanet is announced; however, the current detection techniques cannot say if the planet is truly Earthlike. For that, we ultimately need to get a spectrum, an impression of the planet's atmosphere, to find out whether the planet is indeed habitable. An exoplanet's atmospheric molecules absorb some of the light emanating from its parent star. These absorption features serve as unique signatures of the type and quantity of molecules present in the exoplanet's atmosphere, as different molecules absorb different colors of light.

Table 1: Number of Potentially Habitable Exoplanets as of August 2017. Source: Planetary Habitability Laboratory, University of Puerto Rico at Arecibo (http://phl.upr.edu/projects / habitable-exoplanets-catalog).

Subterranean (Mars size)	Terrain (Earth size)	Superterranean (Super Earth)	Total
1	21	30	52

By the next decade, scientists expect to use the direct-imaging method to measure an Earthlike planet's atmosphere. This is the only planet discovery technique, out of the four widely used, that is capable of separating the planet's reflected light from its star's much brighter glare. (See appendix A to find out how exoplanets are detected.) For comparison, consider this: Earth is ten billion times fainter than the Sun at visible wavelengths. What would it take for an alien civilization in another solar system to take an image of Earth to perform a spectral analysis of its reflected light to find biosignature gasses such as oxygen, carbon dioxide, methane, ozone, and water vapor in our atmosphere? A civilization would surely need very powerful telescopes in space with special devices to filter out Earth's reflected light from the Sun's much brighter shine.

To make the direct-imaging method economically feasible, scientists have proposed to align the James Webb Space Telescope (JWST), which is scheduled to be launched in 2018, with a specially shaped giant screen called an external ocular to capture just the planet's reflected light by blocking its star's light from reaching the telescope's mirror. The ocular will be deployed thirty thousand miles away from JWST and perfectly aligned with it, a difficult but very much achievable proposition.

Today, NASA's Kepler space telescope has provided a critical tally of exoplanets. It has also found a multitude of small, Earthlike exoplanets. But those are too distant from the Earth for any realistic studies of their atmospheres. In the near term, NASA has plans for a new satellite mission called Transiting Exoplanet Survey Satellite (TESS) to be launched in orbit around 2017 that will be capable of finding Earth-size and super-Earth-size exoplanets (up to 1.75 times Earth's size) transiting M stars—stars that are significantly smaller, cooler, and more common than our Sun. The TESS–JWST combination is eventually expected to reveal information on the birth of galaxies, possibly locate habitable worlds orbiting other stars, and track asteroids that might impact Earth. Specifically, TESS will be used to identify the rocky planets and the JWST to observe the planetary atmospheres during their transits or during secondary eclipses. If everything goes according to plan, scientists should be able to infer signs of life on those planets. Finding signatures of life outside our solar system in the next decades may not be so farfetched after all.

Seager's Equation

Recent observations by planetary probes together with ground- and space-based telescopes have shown that water is commonplace throughout our solar system and in the Milky Way galaxy. Oceans of liquid water were detected beneath the icy crusts of two of Jupiter's moons, Europa and Ganymede. Liquid water was also detected on Saturn's moons Enceladus and Titan. Water appears to be in direct contact with the rocky seabed on all these moons. Billions of years ago, Mars was also covered by oceans. The seasonal dark streaks observed today on the Red Planet's surface may be caused by salty flowing water. All this raises the possibility that complex chemical reactions that eventually spawn life are in play.

Observations by NASA's Kepler space telescope suggest that the majority of stars in the sky have planets, many of which may be habitable. Indeed, Kepler's data has shown that rocky worlds like Earth and Mars are probably more abundant throughout the galaxy than gas giants such as Saturn and Jupiter. Many more await discovery in the coming years. And that's just in our Milky Way galaxy. Who knows? Some of them could even harbor intelligent life awaiting discovery as our detection techniques and scientific thinking progress with time.

The goal of exoplanet hunters such as Sara Seager of the Massachusetts Institute of Technology is to detect temperate, Earthlike exoplanets that are appropriate for atmospheric characterization. Inspired by Drake's equation, Seager has recently proposed her own equation to focus simply on the question of whether any alien life is present, not necessarily the technologically advanced kind. The Seager equation looks like this:

$$N = N_* \cdot F_Q \cdot F_{HZ} \cdot F_O \cdot F_L \cdot F_S$$

N = the number of planets with detectable signs of life
N_* = the number of stars observed
F_Q = the fraction of stars that are quiet
F_{HZ} = the fraction of stars with rocky planets in the habitable zone
F_O = the fraction of those planets that can be observed
F_L = the fraction that have life
F_S = the fraction on which life produces a detectable signature gas

Instead of trying to assess the chances of finding radio-capable civilizations, as Drake's equation was designed to do, Seager's equation evaluates the chances of detecting signs of life on exoplanets by signs of biosignature gases. According to her calculations, two inhabited planets could reasonably come to light during the next decade.

To quote Sara Seager: "Someday, sooner or later, we will know of bright stars that host living planets very much like Earth. We will be able to stand beneath a dark sky and point out to our friends or family 'That star has a planet like Earth.' This is a fantastic time for exoplanets, and for astronomy. But the future will be better still" (Seager 2009, 36).

THE SEVEN WORLDS OF TRAPPIST-1

Imagine you are visiting a distant planet where daylight has a salmon hue but the intensity is much less than what you would normally encounter on Earth. You are warm nevertheless because the planet belongs to a red dwarf star, which looms large in the sky, almost three times as big as our Sun. You watch other planets crisscross the sky in the ruddy, perpetual twilight, some of them similar to the Moon in our Earthly sky and some even bigger! Behind you, gleaming in the star's reddish glow, lies the deep ocean that covers a large part of this mystic planet.

Interestingly, this is how the world would hypothetically appear on a planet called TRAPPIST-1*f*. The scenario is quite plausible, as TRAPPIST-1*f* is a member of a seven-planet system belonging to a star—dubbed TRAPPIST-1—located about forty light-years away from us in the constellation Aquarius. TRAPPIST-1 is a red dwarf star whose mass is just 8 percent that of the Sun. Red dwarfs, also called M-dwarfs, are the most abundant stars in the galaxy. They are too faint to be spotted with the naked eye, but their reduced size and limited radiance have an important consequence. They are the longest-lived stars of the galaxy, with lifetimes far greater than that of our Sun. The reason they are so long lived is that they consume their supply of hydrogen fuel at a much slower pace than more massive stars such as our Sun do.

In May 2016, an international team of researchers using TRAPPIST (Transiting Planets and Planetesimals Small Telescope) in Chile reported the discovery of three Earth-size planets orbiting this Jupiter-size star. An ensuing detailed photometric monitoring of the star (i.e., astronomers measured the dipping of the star's light when each planet passed or transited in front of its face) with the global network of ground-based TRAPPIST and space-based Spitzer telescopes revealed four more planets orbiting this star—a total of seven—whose sizes and masses are similar to Earth. Based on their densities, all the TRAPPIST-1 planets are likely to be rocky. The planets, named TRAPPIST-1*b* through *h* (figure 6-2), huddle very close to their host star in tightly separated orbits. The orbital periods of the six inner planets are only 1.51, 2.42, 4.04, 6.06, 9.1, and 12.35 days, while that of the outermost planet is uncorroborated. Computer models suggest that the planets could have formed farther from the star and migrated inward at some point in their cosmic histories.

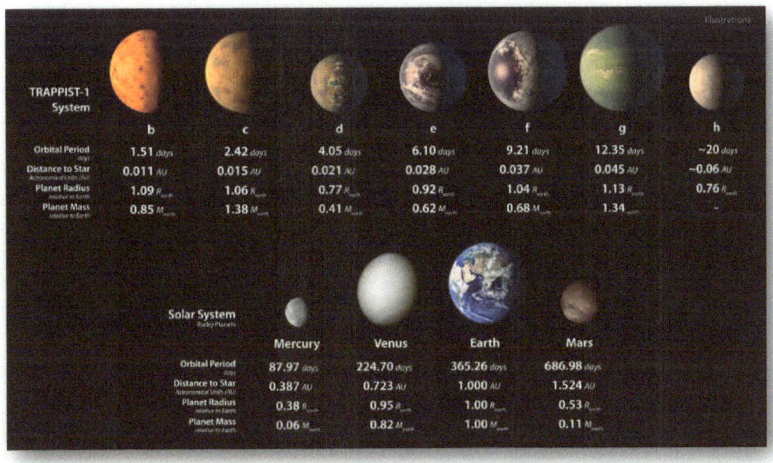

Figure 6-2: Seven Earth-size planets around a tiny, cold red dwarf star called TRAPPIST-1. Three of these planets, TRAPPIST-1e, f, and g, are firmly in the habitable zone. Credits: NASA

The discovery was deemed very newsworthy because of its implications in the ongoing search for extraterrestrial life, and NASA decided to announce the findings at a news briefing at their headquarters in Washington. The results appeared earlier in the journal *Nature* on May 2, 2016.

Can the seven worlds of TRAPPIST-1 support habitable ecosystems? It is too early to tell, but there is a chance that some of them could. The innermost planets, TRAPPIST-1*b*, *c*, and *d*, are probably too hot to support liquid water, although a small amount could exist somewhere on their surfaces, while TRAPPIST-1*h* is too distant and cold to harbor liquid water. TRAPPIST-1*e*, *f*, and *g* are the most interesting planets, as their orbits are confined within the habitable zone—the region where oceans of liquid water could exist on their surfaces, and thus life would be most likely to evolve.

Planets of red dwarf stars were once considered unfit for life to thrive because the habitable zones of red dwarfs overlap their radiation belts, where massive flares of x-rays and UV radiation are commonplace. The flares are the death knell of life as we know it. Besides, some of the closest planets could be tidally locked—the side constantly facing the parent star turning extremely hot, while the far side remaining extremely cold. But new computer models show that some planets in this configuration could still harbor life, provided their atmospheres could dissipate heat across the surfaces. This is a significant result, as it greatly bolsters the chances of the evolution of life in the universe—red dwarfs make up more than three-quarters of the stars in the galaxy.

Curiously, the discovery of the seven worlds of TRAPPIST-1 in some way has vindicated the long-standing imagination of many science-fiction authors who have depicted solar systems comprising spacefaring civilizations with multiple Earthlike planets conveniently suitable for human settlement. However, there is no guarantee that these planets could support life. Yet how cool will it be if we eventually find out that a couple of them really do? Paying a visit is out of question with current rocket technology because they are too far from us. The idea is to scientifically study these planets by remotely observing them with powerful telescopes. This is precisely the goal of the next phase of research: study the atmospheric composition of the planets, determine whether liquid water truly exists on their surfaces, and search for possible signs of life-supporting atmospheres using the Hubble Space Telescope. Hubble's successor—the James Webb Space Telescope, scheduled to be launched in 2018—will be ideally suited to provide a closer look at their atmospheric chemistry. If these planets indeed have atmospheres, they really are the best places to look for life.

Detection of the TRAPPIST-1 Planets

An orbiting planet passing in front of its parent star blocks a small fraction of the starlight in a periodic fashion. The resulting dip in brightness can be measured if the planet's orbits are perfectly aligned with the astronomers' line of vision. (See appendix A1.) Figure 6-3 shows the characteristic dips in TRAPPIST-1's relative brightness with each curve corresponding to a bona fide planet. Information regarding the size of a planet can also be derived from the extent of the unique periodic dips—larger planets block more starlight; hence, the dip in brightness is larger.

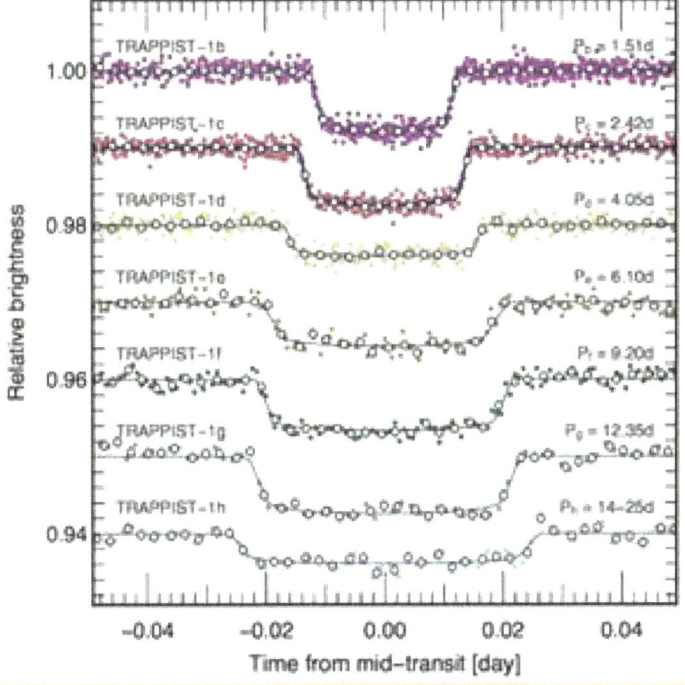

Figure 6-3: Photometric measurements on TRAPPIST-1 resulting from observations by the Spitzer Space Telescope. Credit: M. Gillon et al. (2016, 221–24).

Figure 6-4 shows the representation of the orbits of each of the seven planets where the color codes match each panel. The gray annulus and the two dashed lines are the results of model calculations; they represent the zone around TRAPPIST-1 wherein a planet could potentially harbor oceans of liquid water.

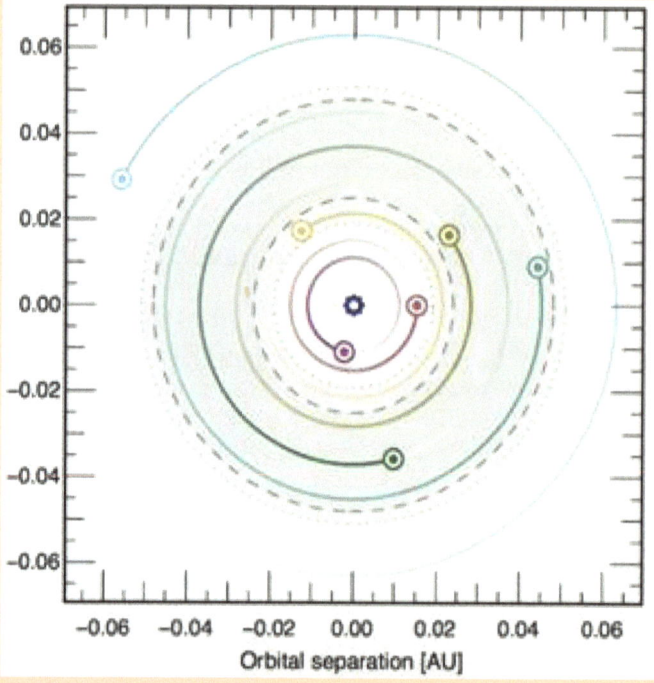

Figure 6-4: The TRAPPIST-1 system as seen by the Spitzer Space Telescope. Credit: M. Gillon et al. (2016, 221–24).

In figure 6-5, the planetary radii are plotted against incident flux or stellar radiation. (The solar radiation on Earth is 1 AU.) Venus and Earth are shown as gray circles; and Mercury, Mars, and Ceres are shown as dotted vertical lines for comparison. The irradiation of the planet TRAPPIST-1*b* shows large errors because its orbital period is unknown.

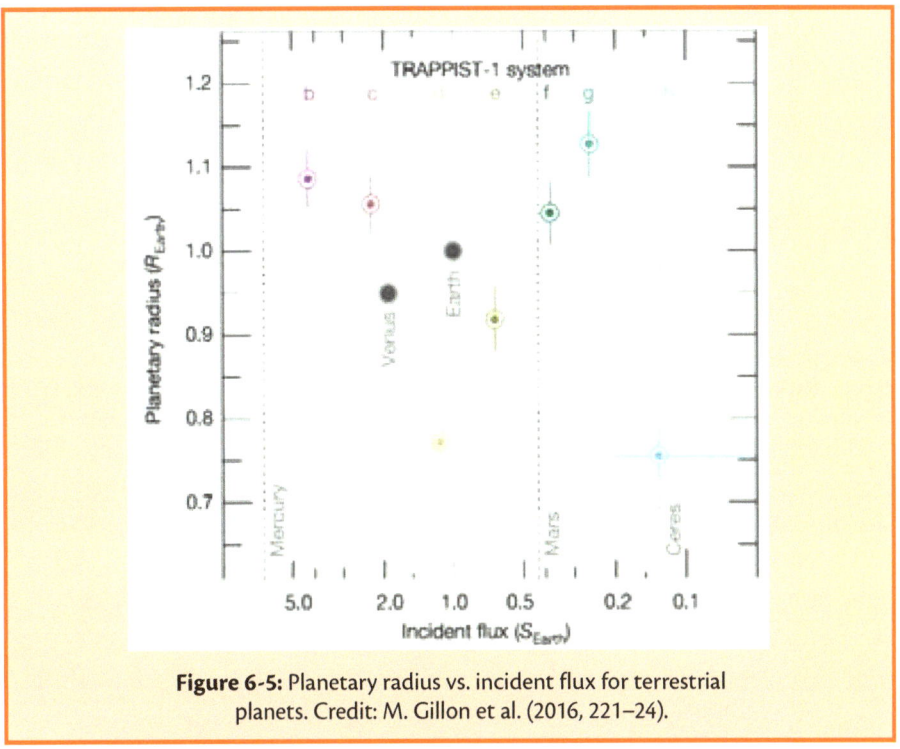

Figure 6-5: Planetary radius vs. incident flux for terrestrial planets. Credit: M. Gillon et al. (2016, 221–24).

Astrophysicist and author Ethan Siegel has beautifully summarized the above ideas in the following lines: "While the discovery of these seven planets is remarkable, the most interesting part of this story is yet to be written. As our telescopes become larger and more sophisticated, we'll finally gain the ability to measure the spectra of these worlds' atmospheres, search for signs of water and life, and perhaps even discover an answer to whether we're not alone in the Universe as far as life goes. With three strong candidates, we might finally need to face the possibility that most of the life that arises in the Universe might occur around stars that aren't like the Sun!" (Siegel 2017).

Interestingly, the TRAPPIST-1 star system will probably outlive our solar system because M-dwarf stars are extremely long lived. When our Sun dies, TRAPPIST-1 will still be a relatively young star and may go on to shine for another trillion years. If there is another part of the universe destined for life to persist, it could very well be in the solar system of TRAPPIST-1, a red dwarf star located in the constellation Aquarius, forty light-years away.

Interstellar Travel

With the discovery of exoplanets, the question that naturally comes to mind is, "Will we be able to visit some of these exotic worlds one day?" Scientists and astronomers do not expect this to happen anytime soon because of the enormous distances that separate us from the stars. According to special relativity, no usable information can travel faster than the speed of light, and hence it would take centuries, even millennia, to travel between stars. Today's fastest rocket would take thirty thousand years to reach the Alpha Centauri, which is the star system closest to us at 4.37 light-years away.

However, there is real hope now of sending unmanned probes that could transmit pictures and other information back to Earth within a generation.

A team of astronomers working with the European Southern Observatory's (ESO) 3.6-meter telescope at La Silla, Chile, along with other telescopes around the world, reported discovery of a planet orbiting our Sun's nearest neighbor, Proxima Centauri. Proxima Centauri is a red dwarf star located in the constellation Centaurus and only 4.23 light-years away from the Sun. Its planet, dubbed Proxima b (Anglada-Escude et al. 2016, 437–440) is at least 1.3 times the mass of Earth and orbits its parent star in 11.2 days, maintaining an average distance twenty times closer than Earth is to the Sun. Despite the close proximity to its parent star, Proxima b actually rides in the crucial habitable zone. This is because Proxima Centauri is cooler, with luminosity only 0.15 percent that of the Sun, which means that its planet is warm and can probably support liquid water on its surface. But Proxima b is likely to be tidally locked with its parent star so that one of its sides perpetually faces its star, just as the Moon does to Earth. And as a result of its closeness to its parent star, Proxima b could be blasted with stellar flares, which are detrimental to life evolving.

Nevertheless, the discovery of Proxima b has opened up the thrilling possibility that one day, humankind will be able to send a probe to this exoplanet. In fact, this is what Stephen Hawking's and Yuri Milner's ambitious Breakthrough Starshot project[10] is aiming to do—send nanocraft (ultralight unmanned spacecraft that travels at 20 percent of the speed of light) to our nearest star system, the Alpha Centauri, within a generation.[11] Its mission: "Seek scientific evidence of life beyond Earth, and encourage public debate from a planetary perspective." Light from an Earth-based phased array of lasers will propel the nanocraft attached to light sails to 20 percent of the speed of light. During the quick flyby, the nanocraft will take pictures of Proxima b and make other measurements. The collected data will be beamed back to Earth, arriving in 4.23 years.

Is humanity eternally destined to belong only to Earth? Or will we be able to reach the stars one day and even build colonies elsewhere in our galaxy? Humanity must continue to venture into space for its future. The journey has just begun.

---

[10] The Breakthrough Initiatives were founded in 2015 by Yuri and Julia Milner to explore the universe. They are a $100-million research and engineering program aiming to demonstrate proof of concept for a new technology, enabling ultralight unmanned space light at 20 percent of the speed of light. Stephen Hawking, along with luminaries Lord Martin Rees, Ann Druyan, and Frank Drake, participated in the launch of project Starshot for interstellar space exploration. The program is led by Pete Worden, the former director of NASA's Ames Research Center and advised by a committee of renowned scientists and engineers. The board consists of Stephen Hawking, Yuri Milner, and Mark Zuckerberg.

[11] Alpha Centauri is a three-star system and the Sun's closest stellar neighbor. Alpha Centauri A and Alpha Centauri B are the two main stars that form a binary pair. Their average distance from Earth is about 4.3 light-years. The third star, Proxima Centauri, is about 4.23 light-years away from Earth; thus, it is our closest star other than the Sun.

Exoplanets and SETI

It should be clear by now that the odds of accidentally stumbling onto an alien radio signal are extremely low. It therefore helps to know where to look. Exoplanets, particularly Earthlike rocky worlds, are the most obvious places to look. Unsurprisingly, astronomers have already started to use data from the Kepler space telescope for more targeted listening to radio signals coming from stars known to have planets orbiting them.

The first exoplanet discovered by Kepler is located 180 light-years away in the constellation Pisces. Named HIP 116454b, the exoplanet measures 2.5 times the diameter of Earth and is therefore dubbed a super Earth. HIP 116454b is by no means a habitable planet, with a rapid nine-day orbit around its host star. Due to its proximity to its host, HIP 116454b is probably tidally locked—the huge gravitational force from the host star would bind HIP 116454b into a tidally locked state with one of its hemispheres facing the host star, just like the Moon does to Earth. Harmful ultraviolet radiation and the huge flares of energy from the nearby star would render HIP 116454b completely inhabitable.

Nevertheless, SETI has pointed the Allen Telescope Array (ATA), located at Hat Creek Radio Observatory 290 miles northeast of San Francisco, California, to study HIP 116454b.

But why would SETI bother aiming ATA at an unlikely target?

With sufficient future funding from its donors, SETI's aim is to examine each planetary system found by Kepler, although the highest priority will be the exoplanets inside their host stars' habitable zones.

So far, the ATA has been looking for signals in the 1,000- to 2,250-megahertz range emanating from HIP 116454b and will soon analyze higher frequencies as well. ATA's forty-two radio antennae allow many frequencies concurrently. If intelligent extraterrestrials have evolved to use similar technologies as we do, then perhaps one day we will be able to listen to their transmissions as well. Unfortunately, nothing up to now indicates that the world is transmitting back to us, and therefore, this is, unluckily, another case of a fruitless search.

"It's a SETI transmission from Earth. Pretend we're not home."

The seven worlds of TRAPPIST-1, discussed earlier, provide an ideal backdrop for speculative sci-fi scenarios. Life could have sprung in one world and spread to the others, naturally, via colliding asteroids or technologically competent beings eager to spread their kind to the rest. One could even envision a multiworld empire consisting of a small federation of planets with interplanetary communication links—an ideal hunting ground for SETI researchers.

While there are some indications that these rocky worlds are similar in composition to mother Earth, no direct observations establishing the existence of oceans of liquid water or life-sustaining atmospheres exist today. That's the job waiting to be done by the James Webb Space Telescope, now scheduled for launch in 2018.

Meanwhile, SETI researchers could look for radio signals that would indicate the presence of this interstellar intelligence. And, indeed, the SETI Institute has used its Allen Telescope Array to observe the environs of TRAPPIST-1, scanning through ten billion radio channels in search of signals. No transmissions have been detected so far.

CHAPTER 7

The Current State of SETI

• • •

Transmission is a diplomatic act, an activity that should be undertaken on behalf of all humans.

—THE SETI 2020 REPORT, 2002

SINCE THE EARLY DAYS OF Project Ozma, the science of SETI has progressed greatly. Countless searches have been conducted—I found a list of about eighty searches through 1999 compiled by Jill Tarter of the SETI Institute.

While most SETI sky searches have studied the radio spectrum, some SETI researchers have considered the possibility that alien civilizations might be using powerful lasers for interstellar communications at optical wavelengths. The idea was first suggested by R. N. Schwartz and Charles H. Townes, one of the inventors of the laser, in a 1961 paper—published in the journal *Nature*—entitled "Interstellar and Interplanetary Communication by Optical Masers." (Schwartz and Townes 1961, 205-8).

It is not possible to cover all the major search efforts in one article, but an excellent account can be found in Alan MacRobert's article "SETI Searches Today" (MacRobert 2009). Here I will describe just two of the searches. The first, called project SERENDIP, used an innovative approach that greatly boosted the availability of radio-telescope time for SETI scientists, leading to the generation of enormous quantities of radio data. The second project, called SETI@home, was conceived out of necessity to properly analyze this large volume of radio data.

It turned out that SETI@home was the first volunteer computing project that greatly mobilized the public's interest in SETI research by tapping into the enormous processing power of millions of personal computers around the world. A new open-source middleware system called the Berkeley Open Infrastructure for Network Computing (BOINC) was developed for volunteer and grid computing. Originally created to support SETI@home, BOINC has become a very useful platform for other distributed applications in areas as diverse as mathematics, medicine, molecular biology, climatology, environmental science, and astrophysics.

Project SERENDIP

SERENDIP is the acronym for Search for Extraterrestrial Radio Emissions from Nearby Developed Intelligent Populations. Using the Arecibo L-band Feed Array (ALFA) on the Arecibo radio telescope[12] in Puerto Rico, project SERENDIP is set up to search for potential signatures of extraterrestrial intelligence by scanning a very broad range of radio band frequencies.

The perpetual problem that SETI researchers face is how to get enough radio-telescope time for their admittedly chancy pursuits. SETI researchers at the University of California at Berkeley came up with the idea of piggybacking an extra receiver onto the Arecibo radio telescope without interrupting the telescope's routine work. Although scientists now have access to large amounts of telescope time, SERENDIP

---

[12] The iconic Arecibo Observatory in Puerto Rico is one of the largest radio telescopes in the world. The observatory suffered considerable damage when the category 4 hurricane Maria hit Puerto Rico in September 2017, causing the island to lose electricity and all communications for an extended period. According to initial reports, the hurricane damaged a smaller, twelve-meter dish of the observatory and caused substantial damage to the main dish. About twenty surface tiles were dislodged, and a ninety-six-foot live feed antenna, which helped focus, receive, and transmit radio waves, disintegrated and fell into the main dish, rupturing it in multiple places. The telescope is not expected to be fully operational for some time, which is a setback especially for SETI research.

researchers do not have control over which targets to study and thus cannot conduct follow-up studies to confirm a possible extraterrestrial signal. Nevertheless, SERENDIP uses the world's largest radio telescope to scan a good fraction of the celestial sphere. This means it can sample many billions of stars in the Milky Way and many thousands of background galaxies.

Figure 7-1: The Arecibo Observatory, a radio telescope in Puerto Rico containing the largest curved focusing dish on Earth, making it the world's most sensitive radio telescope. The telescope has featured in many fundamental discoveries in the field of astronomy. It has even made appearances in motion picture and television productions and got a lot of publicity in 1999 when it began to collect data for the SETI@home project.

Potentially interesting signals are run through false-alarm tests. Those that survive are stored. The most important test is whether the signal repeats itself when the same point on the sky is scanned again. Dedicated follow-up observations are conducted at the locations of the best candidate signals.

The most recently deployed SERENDIP spectrometer, SERENDIP V.v, was installed at the Arecibo Observatory in June 2009 and is currently operational.

SETI@home

The giant radio telescopes of today can pick up and record radio emissions with remarkable efficiency. But the SETI scientists often have to work with limited resources and inadequate computer time. They simply do not have the resources to thoroughly analyze all the available data. As a result, there is the risk that a true signal transmitted by intelligent beings light-years away may remain hidden in the data (radio noise) and be completely ignored.

Then in 1994, David Gedye, a Seattle computer scientist, devised a brilliant scheme to create a dynamic "scientific computing ecosystem." He realized that deeper analysis of SETI radio data would potentially be perfect for distributed computing if tens of thousands of volunteers using home computers could be engaged.

Enter SETI@home, a downloadable screen-saver program that resides on a participant's home computer. During the time when the computer is in an idle state, SETI@home fetches data files ("work units") recorded by the SERENDIP radio receiver from the servers at UC Berkeley. The participant's computer analyzes the downloaded work units utilizing the algorithms built into SETI@home. Depending on the computer's CPU and memory, it could take about ten to twenty hours to analyze the data. When finished, the results are sent back to the UC Berkeley servers. Any possible hits in the analysis are flagged. Another chunk of data is subsequently downloaded from the server, and the process continues.

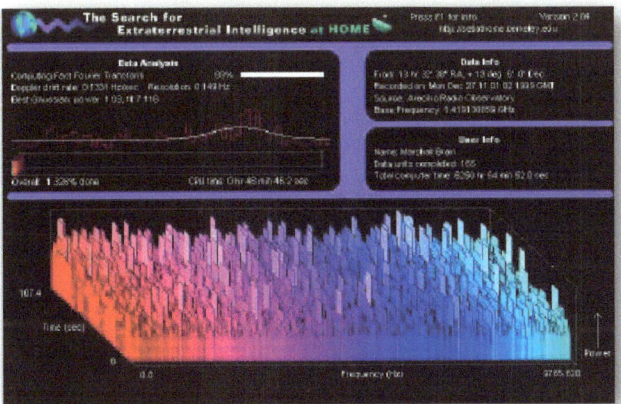

Figure 7-2: SETI@home screen. Sky and Telescope illustration.

SETI@home analyzes only one narrow, 2.5-megahertz segment of SERENDIP's much wider band. The chosen segment is usually centered on the 1,420-megahertz hydrogen line.

Figure 7-2 shows the SETI@home screen. Notice it is divided into three sections: (1) the data-analysis window (upper left), (2) the data and user information (upper right), and (3) the frequency-power-time graph of the data as it is being analyzed (bottom). Data is analyzed using a mathematical technique called a fast Fourier transform (FFT), which is spread out over many channels. A signal of extraterrestrial origin will render itself as a spike that stands out in the background of random equal-shaped signals in all FFT channels. The program flags the information for later analysis by UC Berkeley scientists if the above criterion, and additional ones accounting for the effect of Earth's rotation, are met.

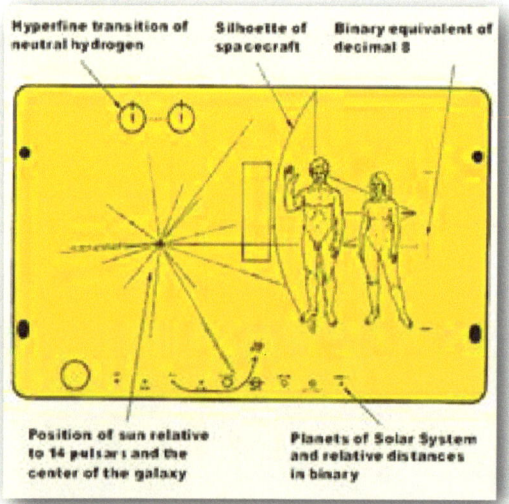

Figure 7-3: The gold plaque placed upon Pioneer 10, the first human artifact to be launched on a trajectory out of the solar system. Image credit: NASA (https://history.nasa.gov/SP-349/epilog.htm).

Today, SETI@home remains the largest computational work effort ever performed. At its peak in 2002, around five hundred thousand people provided nearly one hundred teraflops of computation per second, analyzing radio signals from the Arecibo telescope. SETI@home created an entire new category of distributed computing called volunteer computing that has truly captured the public's imagination. It has helped expand the public's understanding of SETI issues, mobilized enthusiasts, and made them aware of how data is scientifically analyzed.

A Message from Earth

While SETI deals with searching for messages from aliens, the alternative scheme is the attempt to send messages to intelligent extraterrestrial life in the form of radio signals. Messaging to extraterrestrial intelligence or METI (sometimes called active SETI) is a term coined by Russian scientist Alexander Zaitsev. In his paper "Rationale for METI" (Zaitsev 2011), he argues that transmission of the information into the cosmos is one of the pressing needs of an advanced civilization. This view is, of course, not universally accepted, and many believe that the act of transmitting interstellar radio messages is imprudent.

In a new documentary series made for the Discovery Channel, Stephen Hawking said, "We should have been wary of answering back, until we have evolved further." On encountering a more advanced civilization, Hawking says it "might be a bit like the original inhabitants of America meeting Columbus. I don't think they were better off for it."

David Brin, the famous sci-fi author who helped develop the original SETI protocols, has warned of the implications of METI: "Let there be no mistake. METI is a very different thing than passively sifting for signals from the outer space." Carl Sagan, one of the greatest SETI supporters and a deep believer in the notion of altruistic alien civilizations, called such a move "deeply unwise and immature" (Brin 2013 [2006]).

That view is not shared by many people, notably NASA. A few years ago, in 2008, NASA beamed the Beatles' song "Across the Universe" toward the vicinity of Polaris, in the hope that an alien civilization would intercept the transmission. Broadcast over NASA's Deep Space Network, the event commemorated the fortieth anniversary of the day the Beatles recorded the song, as well as the fiftieth anniversary of NASA's founding and the group's beginnings.

The *Pioneer 10* spacecraft, the first man-made object to escape our solar system, carried a plaque (figure 7-3) containing a message for inhabitants of some advanced civilization who might intercept the spacecraft millions of years after the it was launched. The message, devised by Carl Sagan and Jon Lomberg, was intended to communicate the location of the human race, the appearance of an adult male and female of our species, and the approximate era when the probe was launched. A line-drawing of a couple standing in front of the *Pioneer* probe is accompanied by an ingenious scheme for conveying distance, direction, and time information about the spacecraft's origins.

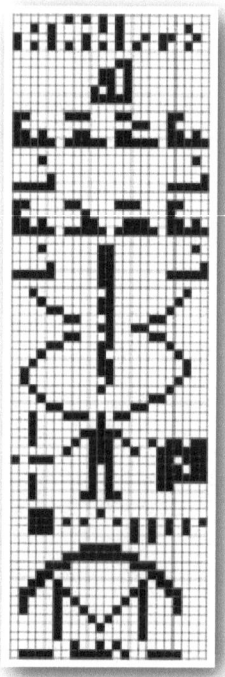

Figure 7-4: A representation of the 1,679-bit Arecibo message.

Even before this, on November 16, 1974, the first message was beamed into space by SETI researchers (Frank Drake, Carl Sagan and colleagues) to celebrate the remodeling of the Arecibo radio telescope. The Arecibo message (figure 7-4) was sent toward the star cluster Messier 13 in the Hercules constellation, which is situated twenty-five thousand light-years away from Earth. The three-minute binary transmission consisting of roughly 210 bytes was composed by Frank Drake, Carl Sagan, and other scientists. It was transmitted using frequency-modulated radio waves.

The Arecibo message consisted of (1) the numbers one through ten; (2) the atomic numbers of the elements that form deoxyribonucleic acid

(DNA); (3) the formulas for the sugars and bases in the nucleotides of DNA; (4) the number of nucleotides in DNA and the double helix structure; (5) a human figure, the physical dimensions of an average man, and the human population of Earth; (6) a representation of the solar system; and (7) a graphic of the Arecibo radio telescope and the physical diameter of the transmitting antenna dish.

The Arecibo message will undoubtedly be seriously degraded by its interaction with cosmic dust in the interstellar medium on its twenty-five-thousand-year-long journey. The loss of even a few bits of information would render the signal undecipherable.

Following the Arecibo transmission, a debate began on the wisdom of attempting communication with extraterrestrial intelligence (CETI), which continues to this day. The US diplomat Michael Michaud suggested that the message had consequences that went beyond science into the realm of politics and proposed a public discussion of the possible benefits and risks of establishing contact. Sir Martin Ryle, the British Nobel Laureate who developed revolutionary radio telescope systems and used them for accurate location of weak radio sources, argued that "any creatures out there [may be] malevolent or hungry." He requested that the International Astronomical Union approve a resolution condemning such attempts at CETI.

In a letter to Sir Ryle, SETI pioneer Frank Drake pointed out, "It's too late to worry about giving ourselves away. The deed is done. And repeated daily with every television transmission, every military radar signal, every spacecraft command[13]...I think that hostile tribes bent on war, be they terrestrial or extraterrestrial, destroy themselves with their own weapons, before they have any notion of how to attempt interstellar travel."

All this did not deter others from attempting to send messages to intelligent extraterrestrial life. Subsequent transmittals had built-in redundancy so that a recipient would be able to reconstruct the message in spite of bit loss due to noise. Some messages had contents based on mathematics

---

[13] Humans have been transmitting strong radio and television programs inadvertently outward from Earth since the 1940s. Our first radio transmissions, traveling at the speed of light, have covered sixty-five light-years from Earth in all directions, reaching more than a thousand star systems.

and physics. The rationale behind that was that any civilization capable of building a device to receive radio waves from outer space must know mathematics and physics to accomplish that task. A list of projects involving messaging to extraterrestrial intelligence can be found in the Wikipedia article entitled "Active SETI."

Interestingly, METI is a scientific effort that has always attracted wide popular curiosity. A high-powered digital radio signal called "Message from Earth" (AMFE) was sent on October 9, 2008, toward the Earthlike extrasolar planet Gliese 581 c, orbiting a red dwarf star, Gliese 581, just twenty-eight light-years away from us. The signal contained 501 messages selected via a competition on the social-networking site Bebo. The message was broadcast using the RT-70 radar telescope of Ukraine's National Space Agency. The signal will reach the planet Gliese 581 c in early 2029. More than half a million people, including celebrities and politicians, participated in the AMFE project, which was the world's first digital time capsule for which the content was selected by the public.

CHAPTER 8

Hints?

• • •

*Imagination will often carry us to worlds that
never were. But without it we go nowhere.*

—Carl Sagan

Clearly, discovery of cosmic intelligence is the Holy Grail of science. As Stephen Hawking pointed out, "There is no bigger question…A Universe full of technological civilizations is a very different place from one with only us." Even the most ardent of skeptics agrees that it is important to know if there is life beyond Earth, and interestingly, nearly two-thirds of Americans believe that some form of alien life exists somewhere in the universe. Although none has been found so far, that does not mean there is no hope at all. Per Hawking[14]:

> They might be there, they might not. But recent experiments like the Kepler mission have changed the game. We now know there are so many worlds, and organic molecules are so common, that it seems quite likely that life is out there, but intelligence is a great unknown. It only took 500 million years to evolve on earth, but it took 2.5 billion years to get from the earliest cells to multi-cell animals, and technological intelligence has appeared only once, so it may be very rare. And when it does evolve, we only need to look in the mirror to know it can be fragile and prone to self-destruction.

---
[14] Quoted in (Fecht 2015).

In the next two sections, we will discuss two recent discoveries in the field of astronomy that have caused much excitement and opened up new areas of research. The phenomena I am about to describe now are most likely produced by perfectly natural astrophysical phenomena—something that no theory has conjured so far. But for the time being, they are unexplained and have aroused much interest among the SETI researchers around the world.

Boyajian's Star

In 2016, the discovery of an extraordinary star, KIC 8462852, was announced. The star was observed by Kepler during its prime mission and first noticed by citizen scientists as part of the Planet Hunters project. Unofficially known as Tabby's star or Boyajian's star, it did something that had never been seen before—its brightness dimmed up to 22 percent at irregular intervals, with variable time scales on the order of days (Boyajian et al. 2016, 3988).

Figure 8-1 is a plot of the full extent of Kepler's light curve data showing the star's relative flux or brightness as a function of time or the number of days since Kepler was launched. The brightness is remarkably constant most of the time. (Relative flux is equal to 1.) However, there are times when multiple large dips in brightness occurred (around day 800 and day 1,500), accompanied by about a dozen small dips throughout the light curve, barely visible at this scale.

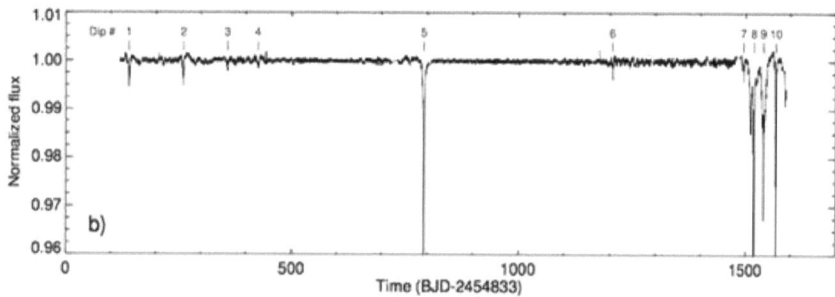

Figure 8-1: The times and extent of ten discrete dips present in the Kepler light curve data.

Further analyses confirmed that the dipping signals in the data were astrophysical in origin; they were not caused by instrumental errors, nor were they artifacts of data processing. Could a black hole[15] be the cause? Unlikely. A nearby black hole's massive gravity would cause the star to wobble, but such characteristic quivering has not been observed. Besides, the black hole would act like a lens (gravitational lensing), actually brightening the light detected by Earth-based telescopes. Spots like sunspots? Unlikely. Spots on the Sun do cause drops in brightness, but the extent of dimming in Boyajian's star is much more than that. Family of light-blocking comets? The comet-swarm hypothesis is plausible for some short-term dips but very unlikely for long-lasting dimming that astronomers witness on Earth.

Recently, a group of astronomers from the Universidad de Antioquia in Colombia suggested that a ringed planet orbiting close to Boyajian's star might be the cause of the enigmatic dipping in the stellar brightness. In order for this to happen, the companion planet must be massively ringed and tilted toward our line of vision to periodically block a substantial fraction of the light emanating from the star. To account for the variable dimming, the researchers suggest that the gravitational interactions between Boyajian's star and its companion planet would make the ring bounce up and down, thus varying the extent of the shading. But the problem with this hypothesis is that precise astronomical measurements should exhibit periodic rather than random dips in the stellar brightness. Moreover, these dips should occur fairly frequently to be noticeable because of the planet's close proximity to its parent star. But no regularly spaced darkening has been spotted yet. The other serious problem is that a massively ringed planet orbiting so close to its parent star cannot be a stable system. The gravitational pull of Boyajian's star would promptly cause the ring to disintegrate—in less than a century of its formation, according to some estimates.

Interestingly, some scientists have attributed the dimming as an artifact of starlight occlusion by an "alien megastructure" such as a Dyson

---

[15] An astronomical object whose immense gravitational field entraps everything, even light, that gets too close (closer than the black hole's event horizon).

sphere orbiting Boyajian's star. But exotic explanations should be the "explanation of last resort," after every other possibility has been ruled out (Wright et al. 2015, 17–38). The problem with an alien explanation is that we do not know how to model the nature of artificial structures or the likelihood of detecting them. Meanwhile, Boyajian's star is receiving a great deal of scrutiny from various observatories around the world because, according to Wikipedia, "It remains an outstanding SETI target because natural explanations have yet to fully explain the dimming phenomenon."

Fast Radio Bursts (FRBs)

In 2007, Duncan Lorimer and his colleagues at West Virginia University discovered an unusual signal buried in the 2001 historical archives of the Parkes radio telescope in Australia (Lorimer et al. 2007, 777–80). The signal was in the form of a five-millisecond radio burst that arrived on August 24, 2001, from an unknown source in deep space, billions of light-years away. But no more subsequent bursts appeared, and the initial excitement faded. However, general acceptance of the Lorimer bursts, also known as the fast radio bursts (FRBs), came after similar bursts were noticed by observers using the Arecibo radio telescope in Puerto Rico (Spitler et al. 2014, 101). FRBs have a characteristic dispersion—the high-frequency waves arrive in the detectors a few hundred milliseconds before the low-frequency ones, just as frequencies change in a slide whistle when the high notes turn into low notes in a few thousandths of a second. Interestingly, the Arecibo signal, called FRB 121102, was found to repeat; its unusual nature prompted astronomers to study it further, using the Karl G. Jansky Very Large Array (VLA) in New Mexico. Eventually, other radio arrays were able to pick up more FRBs from the sky.

So the question is, "What are the FRBs?" The brevity of the signal duration implies that the source is most likely a compact object that discharges an enormous amount of energy—a stellar-mass black hole or a neutron star with a diameter no more than a few hundred kilometers. But nobody knows for sure, and a long list of alternate sources has been proposed. They range from merging black holes to flares on magnetars[16] and even to ETs! (To see why, see box.)

Seth Shostak, the senior astronomer at the SETI Institute, has this to say about FRBs: "…given the strange radio signature of FRBs, it's tempting to wonder if they could be screeches transmitted by intelligent beings. That's not impossible: The raw ingredients for life, including habitable planets, were certainly in place many billions of years ago" (Shostak 2017). He then warns, "You can cook up an alien explanation for just about any sort of signal you discover. Extraterrestrials have been given credit for pulsars, quasars, and lots of other odd celestial behavior. But while extraterrestrials are easy to profile, they are—and should be—hard to convict. There's not been a single case in which aliens were responsible for any new cosmic mystery" (Shostak 2017).

Dispersion Measures of FRBs

The FRBs follow a very interesting pattern that was first noticed by Michael Hippke and his collaborators in a recent publication (Hippke et al. 2015). Each FRB covers a range of radio frequencies, and the dispersion measure (DM) between the arrival time of the high- and low-frequency waves increases with the rising distance covered by the signal. The researchers analyzed the DMs of eleven known FRBs and realized that the DMs are integer multiples of a single number: 187.5 (which is half the lowest DM found, 375 cm pc[17]), appearing in groups centered around 375, 562, 750, 937, 1125 cm pc, with observed errors < 5 percent, which means that there is only a five in ten thousand chance the alignment is coincidental.

---

[16] A type of neutron star with an extremely powerful magnetic field. The magnetic field decay powers the emission of high-energy electromagnetic radiation, particularly x-rays and gamma rays.

[17] A parsec (symbol: pc) is a unit of length used to measure large distances to objects outside our solar system. It is equal to about 3.26 light-years.

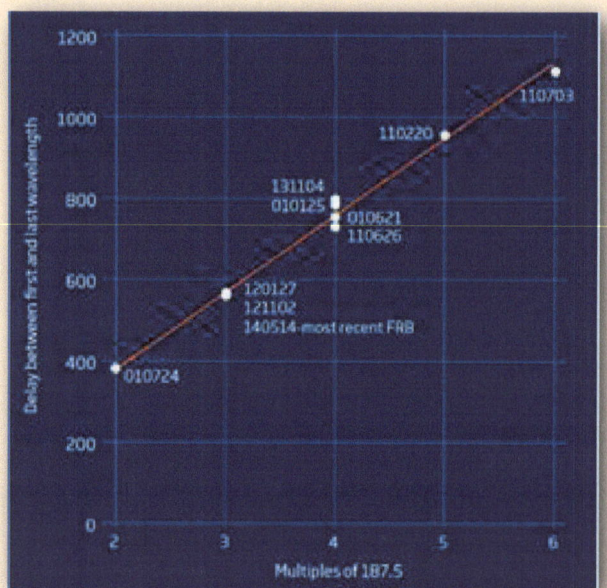

Figure 8-2: DM versus multiple of DMtry =187.5 cm-3 pc for the eleven FRBs. Source: Hippke et al. (2015).

The implications are puzzling. Why are the five sources of the bursts located at regularly spaced intervals from Earth? An extragalactic origin appears unlikely, as intergalactic dust would contribute randomly to the higher DMs. A more plausible explanation would be that the FRBs originated from a source inside our Milky Way (thus, much closer to us), producing the quantized "chirped" signals. This also is odd as it implies the existence of a class of stellar objects that is capable of generating dispersions in discrete steps by a hitherto unknown process. Pulsars do emit bursts of radio waves, but the power is much less and certainly does not follow the regular pattern of the FRBs. The researchers conclude: "If both of these options [extragalactic and galactic origins] could be excluded, only an artificial source (human or non-human) must be considered, particularly since most bursts have been observed in only one location (Parkes radio telescope)." The results, of course, need verification when more FRBs are detected by other telescopes around the world, and it is quite possible that the pattern may completely disappear when more data is accumulated. Nonetheless, the current results are intriguing, particularly from the point of view of SETI.

Are We Alone?

> If aliens are indeed responsible for FRBs, they must possess the technological expertise to produce them because, as pointed out by astronomer Maura McLaughlin of West Virginia University, a co-discoverer of FRBs, it takes a considerable amount of energy to generate a signal with the typical frequency spread seen in the FRBs, particularly if the bursts originate from outside our Milky Way. Extragalactic or not, the energy requirement is undoubtedly huge in both cases, and to generate them, the aliens ought to be much more technologically advanced than we are. In fact, only aliens from a Kardashev Type II civilization (or higher) are speculated to be capable of pulling such a feat.

CHAPTER 9

What Next?

• • •

The Cosmos is all that is or ever was or ever will be.

—Carl Sagan

The story of SETI is unfinished but unfolding nonetheless. No evidence of any extraterrestrial intelligence has been found so far, and every search conducted till now has been fruitless. Complex life is only known to exist on Earth, but scientists are not ruling out other locations in our solar system.

While many people have dedicated their scientific careers to the pursuit of such an elusive goal, their efforts may not bear fruit in their own lifetimes or even in the lifetimes of generations to come. Besides, some scientists and thinkers do not even believe that there are any intelligent species in our galaxy other than our own. So what motivates these contact optimists?

SETI stems from a scientific curiosity, the urge to understand the unknown. There may be no chance of immediate success, but the insatiable yearning to discover is the reason we innovate. This, in my view, is the human spirit—the ability to persevere in the pursuit of truth.

Even the most the ardent skeptics today agree that it is important to know the answer to the question of whether we are alone in the universe. A sensible approach is to build more powerful radio and optical telescopes to improve future directed searches for SETI research and develop state-of-the-art computer algorithms that can efficiently extract an artificially generated signal of extraterrestrial origin from the cosmic dial tone (background noise). It is now possible to scan a much-greater number of frequencies because of the rapid advancement in digital high-bandwidth technology. Some scientists believe there is a small chance that an alien signal may be picked up in the coming decades. It will be an incredible achievement if that really happens. A null result, on the other hand, may simply indicate that more sensitive telescopes capable of detecting even fainter radio flux levels need to be developed.

Scientists do not expect humans to travel to extrasolar planetary systems anytime soon as they are much farther away than the nearest Alpha Centauri star system. Even if we could travel at one-tenth the speed of light, which some futurists think will be possible in a few decades, it would take hundreds to thousands of years to visit them. The obvious question is, "If we cannot go there now, why even look?"

The idea is to scientifically study planets by remotely observing them with the aid of powerful telescopes. The exciting possibility in the immediate future is to predict or actually find the existence of life-forms (not necessarily the intelligent kind) by analyzing the gases in the exoplanet atmosphere (see chapter 6), and if we truly succeed in doing so, humankind will have to revise the way it sees itself in relation to the universe. Perhaps then it will not take its home planet, the "pale blue dot,"[18] for granted anymore. Of the thousands of exoplanets discovered so far, Tony Del Genio, a team member of Cassini's Imaging Science Subsystem (ISS), mused, "It makes me even more excited to think about what surprises other seemingly insignificant 'dots' out there may hold for us if we have the will to search for them, and it emphasizes the commitment we must all make

---

[18] "Pale blue dot" is a reference to Earth made by Carl Sagan, who was a member of NASA's *Voyager 1* mission's imaging team. On February 14, 1990, during the time when the *Voyager 1* spacecraft was preparing to leave our solar system, Sagan convinced the mission scientists to focus its camera on Earth. In the resulting photograph, Earth—appearing as a speck in the vastness of space—was famously labeled by Sagan as "the pale blue dot." Sagan would later write about the photograph in his 1994 book *Pale Blue Dot: A Vision of the Human Future in Space*, inspiring wonderment about the spot that we call home.

to preserve the beauty of the planet that we have for future generations. What Sagan said decades ago is even more true today" (Lindbergh 2017).

As for intelligent aliens, we will have to wait and see.

FURTHER READING

• • •

Cocconi, Giuseppe, and Philip Morrison. 1959. "Searching for Interstellar Communications." *Nature* 184 (4690): 844–46.

The article can also be found online at http://www.coseti.org/morris_0.htm. The modern search for extraterrestrial intelligence was born with the publication of this article in the prestigious journal *Nature*.

Drake, Frank D. 1961. "Project Ozma." *Physics Today* 14 (4): 40–46.

World's first systematic search for alien radio transmissions was conducted more than fifty years ago by Frank Drake, then a twenty-nine-year-old researcher at the National Radio Astronomy Observatory.

Dyson, Freeman John. 1960. "Search for Artificial Stellar Sources of Infrared Radiation." *Science* 131:1667–68.

This is a landmark paper for SETI research, in which the celebrated physicist and futurist Freeman Dyson came up with an entirely new way of looking for signatures of advanced intelligent extraterrestrial life. Dyson himself remarked, "The discovery of an intense point source of infrared radiation would not by itself imply that extraterrestrial intelligence has been found. On the contrary, one of the strongest reasons for conducting a search for such sources is that many new types of natural astronomical objects might be discovered."

Kardashev, Nikolai S. 1964. "Transmission of Information by Extraterrestrial Civilizations." *Soviet Astronomy* 8 (2): 217–21.

Kardashev's ranking of civilizations according to their energy profile can be found in this paper.

Seager, Sara. 2009. *Is there life out there? The search for habitable exoplanets.* Edited by Lee Billings. Distributed from www.saraseager.com.

This exciting, scientifically important article is written for nonspecialists by an expert on exoplanet research. Sara Seager is a professor of planetary science and physics at the Massachusetts Institute of Technology. Online version of this book can be found at: http://seagerexoplanets.mit.edu/ProfSeagerEbook.pdf.

Ward, Peter D., and Donald Brownlee. 2000. *Rare Earth: Why Complex Life is Uncommon in the Universe.* New York, NY: Copernicus Books.

This book makes a counterargument to the principle of mediocrity—the idea that Earth is but one of a myriad of like worlds harboring advanced life. The authors argue that while microbial life may be commonplace in the universe, evolution of complex intelligent life (like the evolution of biological complexity from simple life on Earth) requires an "exceptionally unlikely set of circumstances." Hence, it is likely to be extremely rare at best. While the book has received considerable attention, it is not without its critics as its arguments go against much of the current thinking on the subject. Nevertheless, it could have monumental consequences in the way we perceive at our place in the cosmos if these ideas prove to be correct.

APPENDIX A

• • •

How Exoplanets Are Detected

ADAPTED FROM SARA SEAGER'S ARTICLE "Is There Life Out There? The Search for Habitable Exoplanets" (Seager 2009).

Researchers have developed a handful of techniques to spot planets outside our solar system. These techniques are often used in combination to confirm the initial discovery and learn more about the planet's characteristics. Here is an account of the main methods:

1. **Transits**. When a planet crosses or "transits" the face of its host star, the star's light dims by a small but detectable amount. The probability that any planet's transit will be visible from Earth is low and is dictated by the relative sizes of the star and the planet. Planetary transits are only noticeable for planets whose orbits happen to be perfectly aligned with the astronomers' line of vision. About 10 percent of planets with small orbits have such alignment, and the fraction decreases for planets with larger orbits. Using this

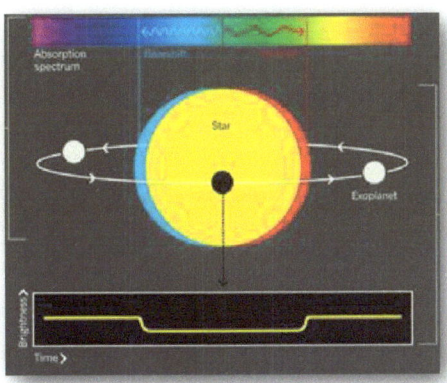

Figure A-1: The brightness of a distant star dips as a planet crosses or "transits" between it and us. Image credit: Sara Seager.

method, researchers can deduce the radius of a planet and its orbital period. Sometimes, a planet's atmosphere can be studied if starlight filters through or reflects off it. This gives information on atmospheric composition, temperature, and cloud formation.

2. **Radial velocity.** The radial-velocity method is currently the most effective method for locating extrasolar planets. Radial velocity is the motion of a star caused by the gravitational influence of its orbiting planets. Radial-velocity measurements can detect only planets whose orbits tug the star toward and away from the observer through the increases (blueshifts) or decreases (redshifts) in the frequency of light the star emits. The exact orbit of an exoplanet is difficult to determine, so radial-velocity measurements let researchers deduce only the time a planet takes to orbit the star (its orbital period) and how its orbit deviates from circular.

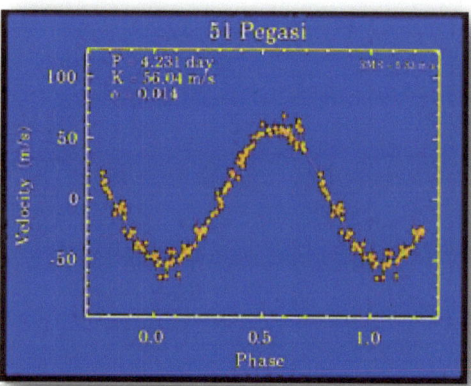

Figure A-2: The radial-velocity graph of 51 Pegasi, the first exoplanet detected. The points on the graph indicate actual measurements taken. The sinusoid is the characteristic shape of the radial-velocity graph of a star wobbling from the tug of an orbiting planet.

Radial velocity is most sensitive for massive planets with short orbital periods. The main drawback is it cannot accurately determine the mass of a distant planet, but it only provides an estimate of its minimum mass.

Scientists can track a star's spectrum using highly sensitive spectrographs that detect periodic shifts of spectral lines toward the red and blue ends of its spectrum. Periodic shifts occurring at fixed intervals of days, months, or even years indicate that the host star's back-and-forth or cyclic motion toward Earth is caused by a body such as a planet orbiting the star.

The radial-velocity method is unlikely to find Earth-size planets that could harbor life. In fact, most of the planets detected by spectroscopy are "hot Jupiters." Cooler planets that are farther away from their host star are much harder to detect with spectroscopy as they produce fewer wobbles in their host star's trajectory and take years to complete a revolution around their host star.

3. **Direct imaging.** A direct image of an exoplanet system is a snapshot of the planets and disk around a central star. Scientists can estimate the orbit of a planet from a time series of images. Direct imaging can provide information about the size, temperature, clouds, atmospheric gases, surface properties, rotation rate, and likelihood of life on a planet from its photometry, colors, and spectra in the visible and infrared regions (Traub and Oppenheimer 2010, 111-156). The technique is appropriate for detecting massive planets that have orbits larger than that of the Neptune in our solar system.

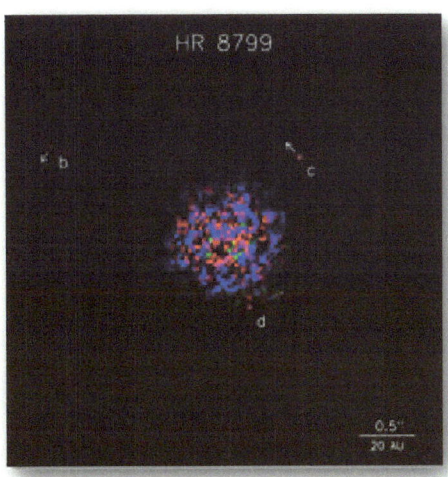

Figure A-3: Three exoplanets orbiting the young star HR 8799. The central star has been suppressed with angular differential imaging, coupled with adaptive optics.

So far, direct imaging has led to the discovery of a handful of exoplanets, including the planets around the stars HR 8799 (figure A-3), Fomalhaut, and Beta Pictoris. The key to the success of direct imaging is the ability to suppress the host star's overwhelming brightness; otherwise, these planets are completely lost in the glare of their host star. For example, if our solar system were viewed from a vantage point seventy light-years away (average for a nearby star), Jupiter would appear roughly a billion times fainter than our Sun. This is equivalent to viewing a dime from a distance of five miles.

Sophisticated instruments (oculars) are being designed for the largest ground-based telescopes that can (1) suppress the host star's image and diffraction pattern, and (2) suppress the star's scattered light from imperfections in the telescope. Studies have shown that direct imaging of exoplanets may become routine using ground-based observatories. Scientists have also proposed aligning a specially shaped occulting screen with the James Webb Space Telescope, to be launched in 2018, for a detailed spectroscopic characterization of planetary mass companions of various stars.

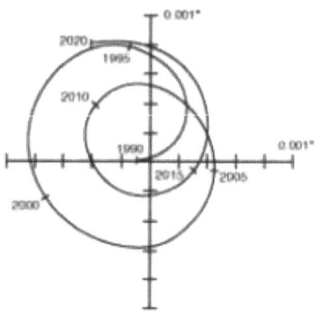

Figure A-4: Astrometric displacement of the Sun due to Jupiter as it would be observed from ten parsecs, or about thirty-three light-years. Image credit: Planet Quest.

4. **Astrometric Method.** Using this method, astronomers look for the motion of the star (wobbles) about the common center of mass. (Both the star and the planet move around a common center of mass.) Astrometric instruments can precisely measure the position of stars as compared to other stars around them and are thus able to detect any movements in the star's position due to the wobbling caused by an orbiting exoplanet. Planets in our solar system have a similar effect on the Sun, producing a to-and-fro motion that can be detected by an observer positioned several light-years away (figure A-4).

APPENDIX B

• • •

New Telescopes

Transiting Exoplanet Survey Satellite

NASA's Transiting Exoplanet Survey Satellite (TESS) mission, scheduled for launch in 2017, will survey nearby stars for transiting exoplanets. Transiting exoplanets are those that pass in front of their parent star as seen from the telescope. This is the same technique NASA's Kepler mission used to discover more than 3,500 exoplanet candidates. TESS will carry four identical specialized wide-field

Figure B-1: A conceptual image of the Transiting Exoplanet Survey Satellite. Image credit: MIT.

CCD cameras, each covering 24 degrees × 24 degrees on the sky with a 100 mm aperture. In a two-year all-sky survey of the solar neighborhood, TESS will cover four hundred times as much sky as Kepler did. In the process, TESS will examine more than a half million bright nearby stars and will likely find thousands of exoplanets with orbital periods (i.e., "years") up to about fifty days. TESS will be capable of finding Earth-size and super-Earth-size exoplanets (up to 1.75 times Earth's size) transiting M stars, stars that are significantly smaller, cooler, and more common than our Sun. TESS is projected to find hundreds of super Earths with a handful of those in an M star's habitable zone. Extensive follow-up observations by ground-based observatories in the United States and internationally will then be used to measure the planet mass to confirm the exoplanets as being rocky.

James Webb Space Telescope

NASA's James Webb Space Telescope (JWST), scheduled to launch in 2018, will be capable of studying the atmospheres of a subset of the TESS rocky exoplanets in visible, near-infrared, and infrared light. The technique JWST will use is called transit spectroscopy. As a transiting exoplanet passes in front of its host star, we can observe the exoplanet's atmosphere as it is backlit by the star. Additional atmospheric observations can be made by observing as the exoplanet disappears and reappears from behind the star. In these observations, the exoplanets and their stars are not spatially separated in the sky but are instead observed in the combined light of the planet–star system. We anticipate TESS will find dozens of super Earths suitable for atmosphere observations by JWST, including several that could potentially be habitable. The chance for life detection with the TESS–JWST combination—albeit small—is a possibility if life turns out to be ubiquitous.

JWST will be stationed at the Second Sun-Earth Lagrange Point, an orbit far beyond Earth's Moon, where gravity will keep it in a fixed orbit. Its giant sunshield will

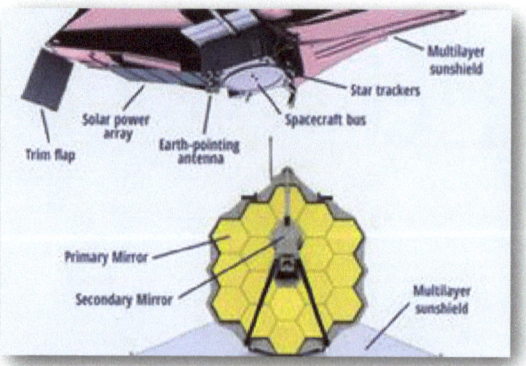

Figure B-2: Artist's conception of the James Webb Space Telescope in space. Image credit: James Webb Space Telescope website.

provide protection from stray heat and light, while its large mirror will enable it to effectively capture the faintest infrared light from early galaxies, newly formed stars, clouds of gas and dust, and much more.

APPENDIX C

• • •

THE TWENTY-ONE-CENTIMETER HYDROGEN LINE

DURING THE SECOND WORLD WAR, a highly significant theoretical work by Grote Reber jump-started the field of radio astronomy. Jan Oort, the director of the Leiden Observatory, Holland (then under Nazi occupation), realized that a spectral line originating from the radio part of the electromagnetic spectrum by atoms or molecules in the interstellar medium (ISM) could, in principle, reveal information about matter distribution at the very early stages of cosmic evolution. About 99 percent of the ISM is gaseous, of which about 90 percent is atomic or molecular hydrogen, and 10 percent helium and traces of other elements.

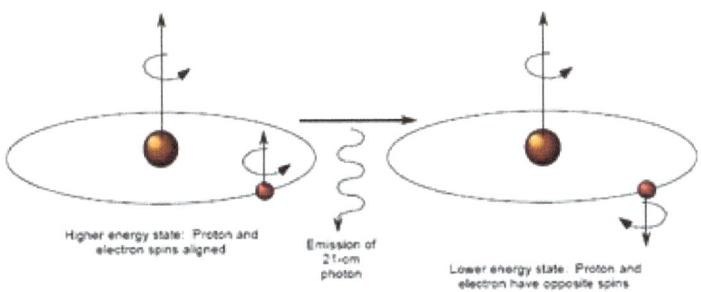

Figure C-1: Illustration of the spin-flip transition in the hydrogen atom, which gives rise to the 21-centimeter line. Credit: http://astronomyonline.org/Science/RadioAstronomy.asp

Oort's student Hendrick van de Hulst subsequently predicted the existence of a spectral line because of a transition between the two spin states of the ground state of hydrogen (figure C-1). This is the famous twenty-one-centimeter line (wavelength = 21.1 centimeters and frequency = 1,420.4 megahertz). It turns out that scientists could draw certain important conclusions about the dynamics of galaxies just by studying the twenty-one-centimeter hydrogen line. The twenty-one-centimeter hydrogen line is a hyperfine transition that arises due to the spin–spin interaction between the electron and the proton in hydrogen: the parallel spin state has a slightly higher energy than that of the antiparallel spin state. Since atoms always try to reside in the lowest possible energy state, an electron in the parallel spin direction will eventually flip to the antiparallel spin direction. This transition is highly forbidden with an extremely low transition probability of $2.6 \times 10^{-15} s^{-1}$. (A hydrogen atom on an average needs to wait a few million years before it undergoes this transition.) Despite its low probability, the twenty-one-centimeter hyperfine transition is one of the main tools of observational astronomy, owing to the very large amount of hydrogen in the universe.

A remarkable feature about the twenty-one-centimeter-line radiation is that it is not blocked by galactic dust! In fact, the twenty-one-centimeter-line radiation provides one of the best ways to map the structure of a galaxy. Most of what is known about the distribution of cold gas in our galaxy, including the mapping of the nearby spiral arms, has come from detailed studies of the variation of twenty-one-centimeter emission across the sky. In 1959, the famous Morrison-Cocconi conjecture concerning the possibility of detecting artificial signals at this wavelength heralded the birth of SETI in its modern form.

REFERENCES

• • •

Anglada-Escude, Guillem, Pedro J. Amado, John Barnes, Zair M. Berdinas, R. Paul Butler, Gavin A. L. Coleman, Ignacio de la Cueva, Stefan Dreizler, Michael Endl, Benjamin Giesers, Sandra V. Jeffers, James S. Jenkins, Hugh R. A. Jones, Marcin Kiraga, Martin Kürster, María J. López-González, Christopher J. Marvin, Nicolás Morales, Julien Morin, Richard P. Nelson, José L. Ortiz, Aviv Ofir, Sijme-Jan Paardekooper, Ansgar Reiners, Eloy Rodríguez, Cristina Rodríguez-López, Luis F. Sarmiento, John P. Strachan, Yiannis Tsapras, Mikko Tuomi, and Mathias Zechmeister. 2016. "A Terrestrial Planet Candidate in a Temperate Orbit around Proxima Centauri." *Nature* 536:437–40.

Braude, S. Y., et al. 2012. "A Brief History of Radio Astronomy in the USSR: A Collection of Scientific Essays." In *The Development of Radio Astronomy at the Sternberg Astronomical Institute of Lomonosov Moscow State University and the Space Research Institute of the USSR Academy of Sciences*, by L. M. Gindilis, edited by S. Y. Braude, B. A. Dubinskii, N. L. Kaidanovskii, N. S. Kardashev, M. M. Kobrin, A. D. Kuzmin, A. P. Molchanov, A. P. Pariiskii, Yu. N. Rzhiga, O. N. Salomonovich, A. E. Samanian, V. A. Shklovskii, I. S. Sorochenko, R. L. Troitskii, V. S., translated by K. I. Kellermann, 97. New York, NY: Springer.

Bell, S. Jocelyn Burnell. 1977. "Petit Four." *Annals of the New York Academy of Science* 302:685–89. http://www.bigear.org/vol1no1/burnell.htm.

Boyajian, T. S., D. M. LaCourse, S. A. Rappaport, D. Fabrycky, D. A. Fischer, D. Gandolfi, G. M. Kennedy, H. Korhonen, M. C. Liu, A. Moor, K. Olah, K. Vida, M. C. Wyatt, W. M. J. Best, J. Brewer, F. Ciesla, B. Csák, H. J. Deeg, T. J. Dupuy, G. Handler, K. Heng, S. B. Howell, S. T. Ishikawa, J. Kovács, T. Kozakis, L. Kriskovics, J. Lehtinen, C. Lintott, S. Lynn, D. Nespral, S. Nikbakhsh, K. Schawinski, J. R. Schmitt, A. M. Smith, G. Y. Szabo, R. Szabo, J. Viuho, J. Wang, A. Weiksnar, M. Bosch, J. L. Connors, S. Goodman, G. Green, A. J. Hoekstra, T. Jebson, K. J. Jek, M. R. Omohundro, H. M. Schwengeler, and A. Szewczyk. 2016. "Planet Hunters IX. KIC 8462852—Where's the Flux?" *Monthly Notices of the Royal Astronomy Society* 457:3988–4004.

Brin, David. n.d. "Shouting At the Cosmos." *Worlds of David Brin, scientist, best-selling author, tech-consultant and speaker.* Accessed October 14, 2017. http://www.davidbrin.com/nonfiction/shouldsetitransmit.html.

Cox, Brian. 2017. "Were We Contacted by Aliens in 1977?" *BBC.* http://www.bbc.co.uk/guides/zqdbgk7#zcgpn39.

Criss, Doug. 2016. "Stephen Hawking Says We've Got about 1,000 Years to Find a New Place to Live." *CNN.* November 18. http://www.cnn.com/2016/11/17/health/hawking-humanity-trnd/index.html.

Darling, David. n.d. "Kardashev Civilizations." http://www.daviddarling.info/encyclopedia/K/Kardashevciv.html.

Darling, David. n.d. "The Worlds of David Darling." http://www.daviddarling.info/encyclopedia/S/SagansResponse.html.

De Sanctis, M. C., E. Ammannito, H. Y. McSween, A. Raponi, S. Marchi, M. Capria, M. T. Capria, F. G. Carrozzo, M. Ciarniello, S. Fonte,

M. Formisano, A. Frigeri, M. Giardino, A. Longobardo, G. Magni, L. A. McFadden, E. Palomba, C. M. Pieters, F. Tosi, F. Zambon, C. A. Raymond, and C. T. Russell. 2017. "Localized Aliphatic Organic Material on the Surface of Ceres." *Science* 355 (6326): 719–22.

Fecht, Sarah. 2015. "Silicon Valley Titan Yuri Milner Bets $100M On Finding Alien Life." *Popular Science*. July 20. https://www.popsci.com/stephen-hawking-yuri-milner-and-frank-drake-announce-worlds-most-extensive-search-alien-life.

Ferris, Timothy. n.d. "Life Beyond Earth." *PBS*. http://www.pbs.org/life-beyondearth/listening/drake.html.

Frank, A., and W. T. Sullivan. 2016. "A New Empirical Constraint on the Prevalence of Technological Species in the Universe." *Astrobiology* 16 (5): 359–62.

Gillon, Michael, Emmanuel Jehin, Susan M. Lederer, Laetitia Delrez, Julien de Wit, Artem Burdanov, Valerie Van Grootel, Adam J. Burgasser, Amaury H. M. J. Triaud, Cyrielle Opitom, Brice-Olivier Demory, Devendra K. Sahu, Daniella Bardalez Gagliuffi, Pierre Magain, and Didier Queloz. 2016. "Temperate Earth-Sized Planets Transiting a Nearby Ultracool Dwarf Star." *Nature* 533 (7602): 221–24.

Gilster, Paul. 2015. *Here's More Evidence That Galactic Super-Civilizations Don't Exist*. 9 22. Accessed 10 15, 2017. https://io9.gizmodo.com/heres-more-evidence-that-galactic-super-civilizations-d-1732390739.

Hippke, M., W. F. Domainko, and J. G. Learned. 2015. "Discrete Steps in Dispersion Measures of Fast Radio Bursts." *ArXiv:1503.05245v2 [astro-ph.HE]*.

Jugaku, J., K. Noguchi, and S. Nishimura. 1995. "A search for Dyson Spheres around late-type stars in the stellar neighborhood." Edited by

G. Seth Shostak. *In Progress in the search for Extraterrestrial Life.* San Francisco: ASP Conference Series. 381-385.

Kaku, Michio. n.d. "The Physics of Extra-Terrestrial Civilizations." *Dr. Michio Kaku. The official web site of famed futurist - physicist - best selling author - Radio and TV personality.* Accessed October 14, 2017. http://mkaku.org/home/articles/the-physics-of-extraterrestrial-civilizations/.

Kaufman, Marc. n.d. "Life, Here and Beyond." *Astrobiology at NASA. Life in the Universe.* https://astrobiology.nasa.gov/about/.

Kennedy, Merrit. 2017. "Scientists Find Signs That Saturn's Moon Enceladus Might Be Hospitable to Life." NPR, April 13. http://www.npr.org/sections/thetwo-way/2017/04/13/523756092/signs-of-hospitality-to-life-found-on-saturns-moon-enceladus.

Lindbergh, Ben. 2017. "'Something Special Is Happening': Cassini's Scientists Honor Their Favorite Photos." The Ringer, September 14. https://www.theringer.com/tech/2017/9/14/16305632/nasa-cassini-mission-scientists-pick-their-favorite-photos.

Lorimer, D. R., M. Bailes, M. A. McLaughlin, D. J. Narkevic, and F. Crawford. 2007. "A Bright Millisecond Radio Burst of Extragalactic Origin." *Science* 318:777–80.

MacRobert, Alan. 2009. "SETI Searches Today." *Sky & Telescope.* March 29. Accessed October 17, 2017. http://www.skyandtelescope.com/astronomy-news/seti-searches-today/.

Sagan, Carl, and William Isaac Newman. 1983. "The Solipsist Approach to Extraterrestrial Intelligence." *Quarterly Journal of the Royal Astronomical Society* 24:113–21.

Schwartz, R. N., and C. H. Townes. 1961. "Interstellar and Interplanetary Communication by Optical Masers." *Nature* 190 (4772): 205-8.

Shostak, Seth. 2017. *What's Causing Those Mysterious 'Bursts' From Deep Space?* January 11. Accessed October 14, 2017. https://www.nbcnews.com/mach/science/cosmic-bruise-could-be-evidence-multiple-universes-ncna771076.

Siegel, Ethan. 2017. "Seven Planets, Including Three Habitable Ones, Found Around Ultra-Cool Dwarf Star." *Forbes*, February 22. https://www.forbes.com/sites/startswithabang/2017/02/22/seven-planets-including-three-habitable-ones-found-around-ultra-cool-dwarf-star/#488c332f2363.

Spitler, L. G., J. M. Cordes, J. W. T. Hessels, D. R. Lorimer, M. A. McLaughlin, S. Chatterjee, F. Crawford, J. S. Deneva, V. M. Kaspi, R. S. Wharton, B. Allen, S. Bogdanov, A. Brazier, F. Camilo, P. C. C. Freire, F. A. Jenet, C. Karako-Argaman, B. Knispel, P. Lazarus, K. J. Lee, J. van Leeuwen, R. Lynch, S. M. Ransom, P. Scholz, X. Siemens, I. H. Stairs, K. Stovall, J. K. Swiggum, A. Venkataraman, W. W. Zhu, C. Aulbert, and H. Fehrmann. 2014. "Fast Radio Burst Discovered in the Arecibo Pulsar ALFA Survey." *The Astrophysical Journal* 790:101–9.

Traub, W. A., and B. R. Oppenheimer. 2010. *Direct Imaging of Exoplanets.*

Edited by S. Seager. Tucson, AZ: University of Arizona Press.

Urban, Tim. 2014. *The Fermi Paradox: Where the Hell Are the Other Earths?* May 23. Accessed October 14, 2017. https://gizmodo.com/the-fermi-paradox-where-the-hell-are-the-other-earths-1580345495.

Wright, Jason T., Kimberly M. S. Cartier, Ming Zhao, Daniel Jontof-Hutter, and Eric B. Ford. 2015. "The Ĝ Search for Extraterrestrial Civilizations with Large Energy Supplies. IV. The Signatures and Information Content of Transiting Megastructures." *The Astrophysical Journal* 816:17–38.

Zaitsev, A. L. 2011. "Rationale for METI." *ArXiv* 1105.0910v1.

INDEX

• • •

21 cm hydrogen line, 83
6EQUJ5, 4
accelerator, xvi
Allen Telescope Array (ATA), 52, 53
Alpha Centauri, 50, 51, 71
A Message from Earth (AMFE), 62
Aquarius, 44, 50
Arecibo
    L-band Feed Array (ALFA), 55
    message, 54, 60, 61
    Observatory, 55, 56
    radio telescope, 55, 59, 60, 61, 66
    signal, 58, 66
    transmission, 61
Ariel, 29
artificial intelligence, 15
asteroid, 16, 19, 26, 29, 35–37, 42, 53
Astrometric Method, 78
Australopithecus, 27
Beatles, 60
Bebo, 62
Bell, J., 6, 7, 8
Beta Pictoris, 77

Big Ear, 4, 5
black hole, 6, 65, 66, 67
Berkeley Open Infrastructure for Network Computing (BOINC), 55
Boyajian's Star, 64–66
Breakthrough Starshot, 51
Brin, David, 59
Brownlee, Donald, 26
Butler, Paul, 39
Callisto, 29, 30
Carus, Titus Lucretius, xv
Cassini
    probe, 33, 34
    spacecraft, 35
    Imaging Science Subsystem (ISS), 71
Centaurus, 50
communication with extraterrestrial intelligence (CETI), 61
chaos terrains, 31
Charon, 29
Chi Sagittarii, 5

Churyumov-Gerasimenko, xv, 36
civilization
Type 0, 21
Type 1.1, 21
Type I, 20–22
Type II, 20–22, 69
Type III, 20–22
Clarke, Arthur C., viii, xvii
Cocconi, G., 1, 2, 4, 84
Cosmological principle, 24
Cox, Brian, 19
Curiosity, xviii, xiv
Darling, David, 20, 21, 24
Dawn, 36, 37
Del Genio, Tony, 71
Democritus, xiv
direct imaging, 41, 42, 77
Drake equation, 12–17, 19, 43
Drake, Frank, 3, 13–16, 51, 60, 61, 73
Druyan, Ann, 51
Dyson
ring, 11, 18
Sphere, 9–12, 20, 21, 65
swarm, 10, 11
Dyson, Freeman, 8–10, 20, 21, 73
Ehman, J., 4
Enceladus, 29, 33–35, 37, 42
Eris, 29
European Southern Observatory (ESO), 50
Siegel, Ethan, 49
Europa, xv, 29–33, 35, 37, 42

Europa Clipper, 32, 33, 35
existential risk, xvi, 15
exoplanet, 16, 38–44, 50–52, 71, 74–80
extraterrestrial
civilization, *xiv*, 19, 28, 74
intelligence, *xiv*, *xvi*, 1, 11, 17, 27, 30, 52, 55, 59, 61, 62, 70, 73
life, *xiv*, 45, 59, 61, 73
origin, 58, 71
radio emmission, 55
radio signal, 20
signal, 56
Fast Fourier Transform (FFT), 58
Fermi paradox, 17–19, 27
Fermi, E., 17, 19
Fomalhaut, 77
Frank, A., 16, 17
fast radio bursts (FRB), 66–69
Galilean moons, 30
Galilee, Galileo, 30
Galileo mission, 31
Ganymede, 29, 30, 37, 42
Gedye, David, 57
Gliese 581, 62
Gold, T., 8
Goldilocks zone, 40, See also habitable zone
Goldsmith, Don, 10
Great Silence, 19
habitable zone, 17, 26, 40, 41, 43, 45, 46, 50, 52, 80

Hat Creek Radio Observatory, 52
Hawking, Stephen, xiv, 15, 16, 51, 59, 63
Hercules, 60
Hewish, Antony, 6–8
HIP 116454b, 52
Hippke, Michael, 67, 68
Hoyle, Fred, 8
HR 8799, 77
Hubble Space Telescope, xv, 31, 32, 46
Huygens, Christiaan, xv
hydrogen line, 4–6, 58, 84
hydrothermal activity, 31, 33
hydrothermal vent, 31, 34
Iapetus, 29
IBM, 4
Jansky, Karl G., 66
Jupiter, xv, 26, 30–33, 36, 37, 39, 40, 42–44, 77, 78
James Webb Space Telescope (JWST), 35, 42, 46, 53, 78, 80
Kaku, Michio, 18, 19, 25, 38
Kardashev, Nikolai, 20, 21, 22, 74
Kardashev's scale, 21
Kepler
    mission, 39, 63, 79
    satellite, 17
    space telescope, 42, 43, 52, 64
Kepler, Johannes, xv
KIC 8462852, 64
LGM, 7

Lomberg, Jon, 60
Lorimer, Duncan, 66
Lowell, Percival, xv
M-dwarf, 44, 49
M star, 80
MacRobert, Alan, 54
magnetar, 67
Marcy, Geoffrey W., 39
Mars, *xiii–xvi*, 36, 37, 41, 43, 48, *See also* red planet
Mars Orbiter, *xiv*
Mayor, Michel, 39
Mayr, Ernst, 27
Messier 13, 60
methane, *viii*, *xv*, 34, 41
methanogenesis, 34
methanogens, 34
Messaging to extraterrestrial intelligence (METI), 59, 62
Michaud, Michael, 61
Milky Way, *v*, *vi*, *xvi*, 16, 18, 25, 39, 42, 43, 56, 68, 69
Milner, Yuri, 51
Miocene epoch, 27
Morrison, Philip, 1, 2, 4, 73
nanocraft, 51
NASA, *xiii*, *xv*, *xvi*, 12, 17, 31–33, 35–37, 39, 42, 43, 45, 51, 58, 60, 71, 79, 80
Neptune, 29, 40, 77
neutron star, 8, 66, 67
Newman, William I., 24, 25, 27, 28

Oberon, 29
Oort, Jan, 83, 84
Orcus, 29
Parkes radio telescope, 66
pessimism line, 17
Pioneer 60
Pluto, 29
Polaris, 60
proconsul, 27
Project Ozma, 3, 54, 73
prokaryotic bacteria, 26
Proxima b, 50, 51
Proxima Centauri, 50, 51
pulsar, 8, 39, 67, 68
quasar, 6, 22, 67
Queloz, Didier, 39
radial velocity, 39, 76, 77
radio telescope, 3–6, 54–57, 60, 61, 66, 70, 71
Rare Earth, 19, 25, 26, 74
Reber, Grote, 83
red dwarf, 44, 46, 50, 62
Red Planet, *xiii, xv*, 37
Rosetta, *xv*, 36
Ryle, Martin, 8, 61
Sagan, Carl, 21, 24, 25, 27, 28, 38, 59, 60, 63, 70–72
Sagittarius, 3, 5
Schwartz, R. N., 54
sci-fi, 53, 59
Seager, Sara, 38, 40, 42, 43, 44, 74, 75
Sedna, 29
SERENDIP, 54–58

SETI, *xiv, xvi, xvii*, 2, 3, 6, 8, 19, 20, 22, 23, 52–57, 59–61, 64, 66–68, 70, 71, 73, 84
Shostak, Shostak, *xiv*, 67
Slysh, Vyacheslav, 12
Sobel, Dava, 3
solar system, *xv*, 6, 7, 26, 29, 30, 33, 35–40, 42, 49, 50, 58, 60, 61, 67, 70, 71, 75, 77, 78
Spitzer Space Telescope, 47
Sullivan, W. T., 16, 17
super Earth, 41, 52
Superconducting Super Collider, *xvi*
Titania, 29
Transiting Exoplanet Survey Satellite (TESS), 42, 79, 80
tidal locking, 31, 46, 50, 52
Tipler, Frank, 24
Titan, 29, 43
Townes, Charles H., 54
Transits, 75
TRAPPIST-1, 44–50, 53
Triton, 29
Umbriel, 29
van de Hulst, Hendrick, 84
Virgo, 39
Very Large Array (VLA), 66
von Neumann probe, 18, 24, 25
von Neumann, John, 18
Voyager, 71
Voytek, Mary, 34, 35
Ward, Peter, 26, 74
water plumes, 31–35

Whitman, Walt, 23
Wide Field Infrared Survey Explorer (WISE), 12
Wolszczan, Alexander, 39
Wow! Signal, 3–6
Zaitsev, Alexander, 59

www.ingramcontent.com/pod-product-compliance
Lightning Source LLC
Chambersburg PA
CBHW040220220526
45473CB00001B/57